PAPUA NEW GUINEA

Mathematics

Essential Skills

4

Pat Lilburn

Contents

For Students

The *Oxford Essential Skills Book* has been written to support the *Oxford Grade 4 Mathematics Student Books A* and *B*.

The focus of this book is on providing practice examples to revise and support the concepts learned in the two Student Books.

The book is set out under the five Strands outlined in the *Lower Primary Mathematics Syllabus* and the *Lower Primary Mathematics Teacher Guide*:

- Number and Application
- Measurement
- Space and Shape
- Chance and Data
- Patterns

Within each Strand, you will find many practice examples that relate directly to the Learning Outcomes for that Strand. These Learning Outcomes are specified in the *Lower Primary Mathematics Teacher Guide*.

The subheading in each Strand shows you what the examples will help you practise.

4.1.1 Count, order, read and record three and four-digit numbers

The Help Box contains worked examples to show you how to set out and complete the examples in each Strand. The Remember Box contains information that will help you complete the examples.

Help Box

Add whole numbers without trading:	Add whole numbers with trading in one or more places:
352 + 433	287 + 576
3 5 2	2 8 7
+ 4 3 3	+ [1]5 [1]7 6
7 8 5	8 6 3

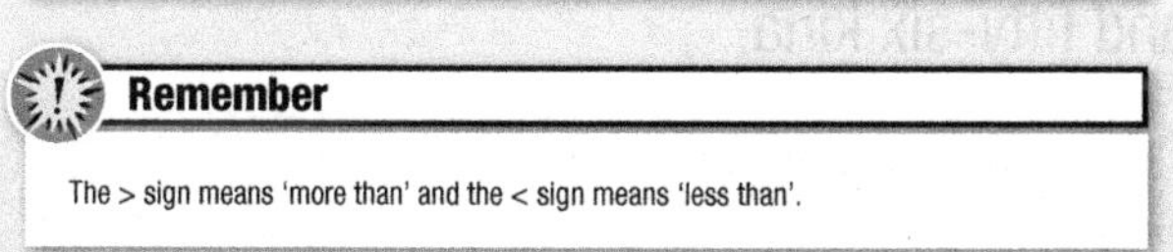

Remember

The > sign means 'more than' and the < sign means 'less than'.

At the end of each Strand there is an Assessment Section that covers all the examples from that Strand. If you have difficulty with any test questions, go back to the relevant page in the Strand and look at the worked examples before doing the test again.

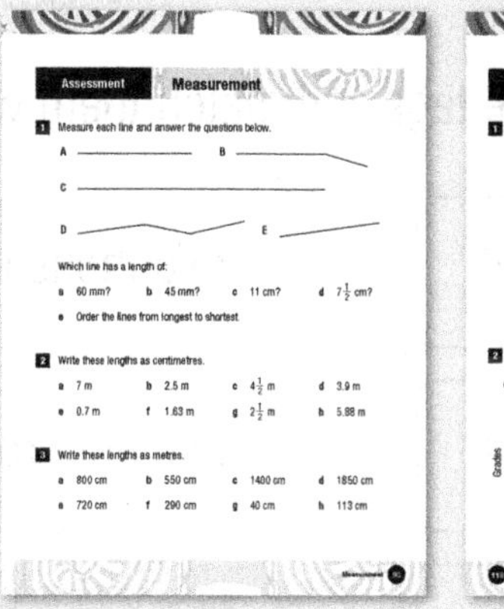

Assessment **Measurement**

1 Measure each line and answer the questions below.

A B C D E

Which line has a length of:

a 60 mm? b 45 mm? c 11 cm? d $7\frac{1}{2}$ cm?

e Order the lines from longest to shortest.

2 Write these lengths as centimetres.

a 7 m b 2.5 m c $4\frac{1}{2}$ m d 3.9 m

e 0.7 m f 1.63 m g $2\frac{1}{2}$ m h 5.88 m

3 Write these lengths as metres.

a 800 cm b 550 cm c 1400 cm d 1850 cm

e 720 cm f 290 cm g 40 cm h 113 cm

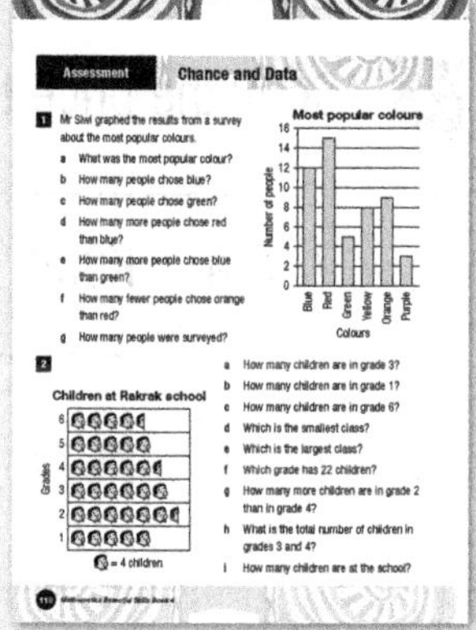

Assessment **Chance and Data**

1 Mr Siwi graphed the results from a survey about the most popular colours.

a What was the most popular colour?

b How many people chose blue?

c How many people chose green?

d How many more people chose red than blue?

e How many more people chose blue than green?

f How many fewer people chose orange than red?

g How many people were surveyed?

2

a How many children are in grade 3?

b How many children are in grade 1?

c How many children are in grade 6?

d Which is the smallest class?

e Which is the largest class?

f Which grade has 22 children?

g How many more children are in grade 2 than in grade 4?

h What is the total number of children in grades 3 and 4?

i How many children are at the school?

There is an Answers section and a section for Important Facts at the back of the book. The Answers section contains answers to all practice examples and test questions while the Important Facts section contains some of the facts that you will need to refer to throughout the year.

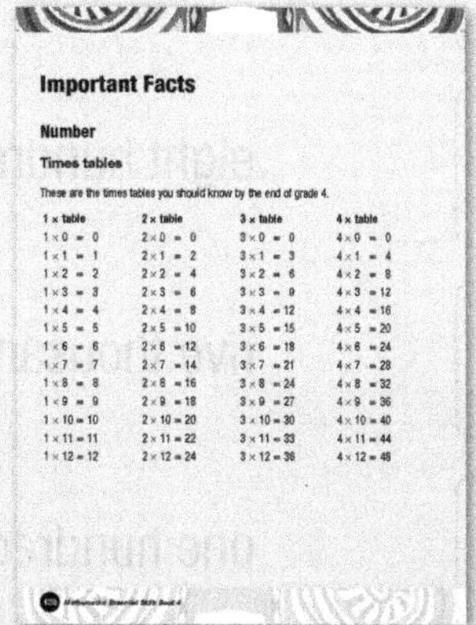

Important Facts

Number

Times tables

These are the times tables you should know by the end of grade 4.

1 × table	2 × table	3 × table	4 × table
1 × 0 = 0	2 × 0 = 0	3 × 0 = 0	4 × 0 = 0
1 × 1 = 1	2 × 1 = 2	3 × 1 = 3	4 × 1 = 4
1 × 2 = 2	2 × 2 = 4	3 × 2 = 6	4 × 2 = 8
1 × 3 = 3	2 × 3 = 6	3 × 3 = 9	4 × 3 = 12
1 × 4 = 4	2 × 4 = 8	3 × 4 = 12	4 × 4 = 16
1 × 5 = 5	2 × 5 = 10	3 × 5 = 15	4 × 5 = 20
1 × 6 = 6	2 × 6 = 12	3 × 6 = 18	4 × 6 = 24
1 × 7 = 7	2 × 7 = 14	3 × 7 = 21	4 × 7 = 28
1 × 8 = 8	2 × 8 = 16	3 × 8 = 24	4 × 8 = 32
1 × 9 = 9	2 × 9 = 18	3 × 9 = 27	4 × 9 = 36
1 × 10 = 10	2 × 10 = 20	3 × 10 = 30	4 × 10 = 40
1 × 11 = 11	2 × 11 = 22	3 × 11 = 33	4 × 11 = 44
1 × 12 = 12	2 × 12 = 24	3 × 12 = 36	4 × 12 = 48

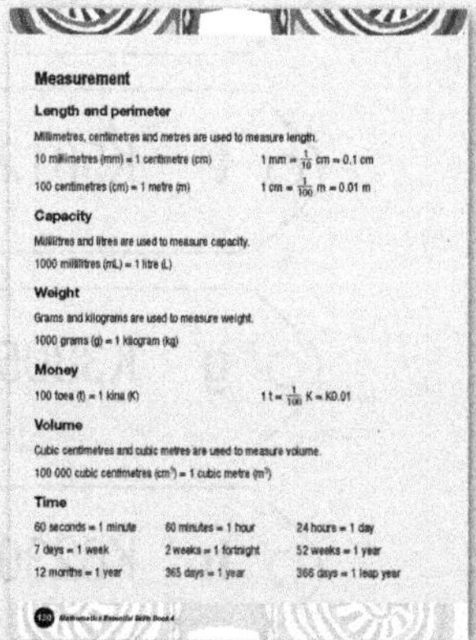

Measurement

Length and perimeter

Millimetres, centimetres and metres are used to measure length.

10 millimetres (mm) = 1 centimetre (cm) $1 \text{ mm} = \frac{1}{10} \text{ cm} = 0.1 \text{ cm}$

100 centimetres (cm) = 1 metre (m) $1 \text{ cm} = \frac{1}{100} \text{ m} = 0.01 \text{ m}$

Capacity

Millilitres and litres are used to measure capacity.

1000 millilitres (mL) = 1 litre (L)

Weight

Grams and kilograms are used to measure weight.

1000 grams (g) = 1 kilogram (kg)

Money

100 toea (t) = 1 kina (K) $1 \text{ t} = \frac{1}{100} \text{ K} = \text{K}0.01$

Volume

Cubic centimetres and cubic metres are used to measure volume.

100 000 cubic centimetres (cm^3) = 1 cubic metre (m^3)

Time

60 seconds = 1 minute	60 minutes = 1 hour	24 hours = 1 day
7 days = 1 week	2 weeks = 1 fortnight	52 weeks = 1 year
12 months = 1 year	365 days = 1 year	366 days = 1 leap year

Strand Number and Application

Count, order, read and record three and four-digit numbers

Read and write three and four-digit numbers in words and numerals

1 Copy the price tags and match them with their words.

a K809	two thousand, six hundred and sixty-seven kina
b K574	two hundred and ninety kina
c K290	three thousand, eight hundred and five kina
d K156	six thousand, two hundred and thirty kina
e K2667	five hundred and seventy-four kina
f K5012	eight hundred and nine kina
g K3805	five thousand and twelve kina
h K6230	one hundred and fifty-six kina

2 Write these numerals as words.

a	416	**b**	845	**c**	170	**d**	638	**e**	982
f	3805	**g**	4033	**h**	7418	**i**	1961	**j**	6274
k	5410	**l**	9855	**m**	2006	**n**	8724	**o**	1279

3 Write these words as numerals.

a five hundred and seventeen

b two hundred and thirty-five

c eight hundred and forty-one

d four hundred and nine

e one hundred and twenty-eight

f seven hundred and sixty-three

g six thousand, four hundred and seventy-four

h nine thousand, seven hundred and eight

i three thousand and eighty-five

j eight thousand, five hundred and nineteen

k two thousand, six hundred and fifty-seven

l seven thousand, six hundred and forty-two

4 Write in numerals the number that is ten more than these numbers.

- **a** eight hundred and sixty-five
- **b** three hundred and fifty
- **c** one hundred and thirteen
- **d** six hundred and twenty-three
- **e** four hundred and eighty-one
- **f** five hundred and seven
- **g** one thousand and thirty-seven
- **h** nine thousand and fifteen
- **i** five thousand, two hundred and fifty-six
- **j** two thousand, five hundred and forty-nine
- **k** nine thousand, three hundred and twelve
- **l** seven thousand, four hundred and ninety-five

5 Write in words the number that is one hundred more than these numbers.

a 523	**b** 782	**c** 209	**d** 355	**e** 816
f 6782	**g** 2171	**h** 4277	**i** 9550	**j** 1609
k 3128	**l** 8759	**m** 5080	**n** 7836	**o** 9264

6 Write in numerals the number that is ten less than these numbers.

- **a** two hundred and twenty-six
- **b** five hundred and eighty

c seven hundred and ninety-four

d nine hundred and eleven

e four hundred and thirty-one

f one hundred and sixty-nine

g six thousand and fifty-eight

h five thousand and nineteen

i eight thousand, seven hundred and forty-five

j three thousand, one hundred and thirty

k nine thousand, five hundred and seventy-two

l one thousand, three hundred and eighty-eight

Compare three and four-digit numbers

1 Write a number that is larger than each of these numbers.

a 459	**b** 705	**c** 299	**d** 640	**e** 137
f 986	**g** 2775	**h** 8116	**i** 7000	**j** 1999
k 9000	**l** 4500			

2 Write a number that is smaller than each of these numbers.

a 700	**b** 310	**c** 675	**d** 129	**e** 850
f 400	**g** 5862	**h** 2500	**i** 3035	**j** 1000
k 6047	**l** 8006			

Remember

The > sign means 'more than' and the < sign means 'less than'.

3 Use > or < to fill the gap.

a	516 ☐ 561	**b**	890 ☐ 809	**c**	338 ☐ 383
d	293 ☐ 299	**e**	415 ☐ 405	**f**	651 ☐ 615
g	180 ☐ 108	**h**	901 ☐ 910	**i**	737 ☐ 773
j	2300 ☐ 3200	**k**	5060 ☐ 5106	**l**	1500 ☐ 1450
m	6705 ☐ 6570	**n**	4875 ☐ 4758	**o**	7312 ☐ 7321
p	3866 ☐ 3688	**q**	9110 ☐ 9107	**r**	2745 ☐ 2754

4 Write a number to make each number sentence true.

a	675 > ______	**b**	430 < ______	**c**	816 > ______
d	200 > ______	**e**	______ > 599	**f**	______ < 320
g	899 < ______	**h**	106 > ______	**i**	773 > ______
j	______ > 3990	**k**	______ > 5000	**l**	______ < 9660
m	2113 > ______	**n**	6080 < ______	**o**	3686 < ______
p	______ < 4800	**q**	______ > 7019	**r**	______ > 5949
s	7399 < ______	**t**	1701 > ______	**u**	6404 > ______

Order three and four-digit numbers

1 Write the largest number in each group.

a 750 570 757 575

b 486 648 684 468

c 919 909 910 109

d 530 305 503 350

e 274 747 724 740

f 310 130 301 103

g 516 156 561 165

h 938 983 893 839

i 626 622 662 620

j 550 515 551 505

k 812 821 728 877

l 476 746 740 674

m 138 183 108 180

n 239 392 329 293

o 711 771 717 770

p 689 698 869 690

q 354 350 335 345

r 927 970 972 907

s 556 565 560 550

t 817 870 807 810

u 253 235 233 252

v 145 154 144 155

w 408 480 481 400

x 756 765 657 576

y 963 936 930 960

z 338 330 383 380

2 Write the smallest number in each group.

a	4255	4525	4552	4225	**b**	6719	6971	6179	6197
c	2583	2385	2358	2835	**d**	1074	1047	1407	1040
e	7296	7926	7269	7629	**f**	5883	5838	5833	5888
g	5409	5490	5410	5904	**h**	3681	3861	3618	3816
i	1925	1295	1259	1529	**j**	8014	8140	8041	8040
k	2660	2066	2060	2606	**l**	4797	4779	4977	4997
m	9652	9256	9562	9265	**n**	6184	6048	6148	6044
o	3514	3541	3415	3451	**p**	5041	5014	5004	5040
q	7989	7988	7899	7898	**r**	1282	1228	1822	1828
s	4456	4546	4465	4455	**t**	9095	9059	9050	9090
u	5261	5216	5266	5260	**v**	8586	8568	8685	8658
w	2327	2237	2273	2227	**x**	3242	3222	3422	3244

3 Write in order three numbers that come between each pair of numbers.

a	315 and 750	**b**	150 and 500	**c**	400 and 600
d	295 and 410	**e**	430 and 590	**f**	180 and 300
g	350 and 450	**h**	120 and 220	**i**	690 and 750
j	610 and 700	**k**	750 and 800	**l**	880 and 930
m	590 and 615	**n**	695 and 705	**o**	398 and 410

4 Write in order three numbers that come between each pair of numbers.

a 1500 and 2500 **b** 1000 and 1500 **c** 5300 and 6000

d 2500 and 3000 **e** 3500 and 4000 **f** 2700 and 3000

g 7200 and 7500 **h** 6100 and 6200 **i** 3800 and 3900

j 6950 and 7000 **k** 5700 and 5850 **l** 4900 and 4950

m 3050 and 3100 **n** 8360 and 8380 **o** 7450 and 7470

5 Put these amounts of money in order from least to most.

a K4000 K3750 K410 K4150 K800 K3500

b K650 K1670 K965 K1855 K1280 K825

c K2300 K2030 K2160 K260 K1030 K1130

d K525 K5500 K5250 K590 K1550 K5090

e K8950 K895 K8075 K8070 K870 K8700

f K3100 K3350 K1350 K3175 K1730 K3085

g K775 K7050 K7500 K5700 K707 K7110

h K9880 K890 K9080 K908 K8990 K8090

i K4600 K4060 K460 K4160 K466 K4660

j K995 K1090 K1019 K990 K1990 K1190

Represent three and four-digit numbers

1 If [thousand cube] = 1000, [hundred flat] = 100, [ten rod] = 10 and [unit cube] = 1, write the numbers shown below.

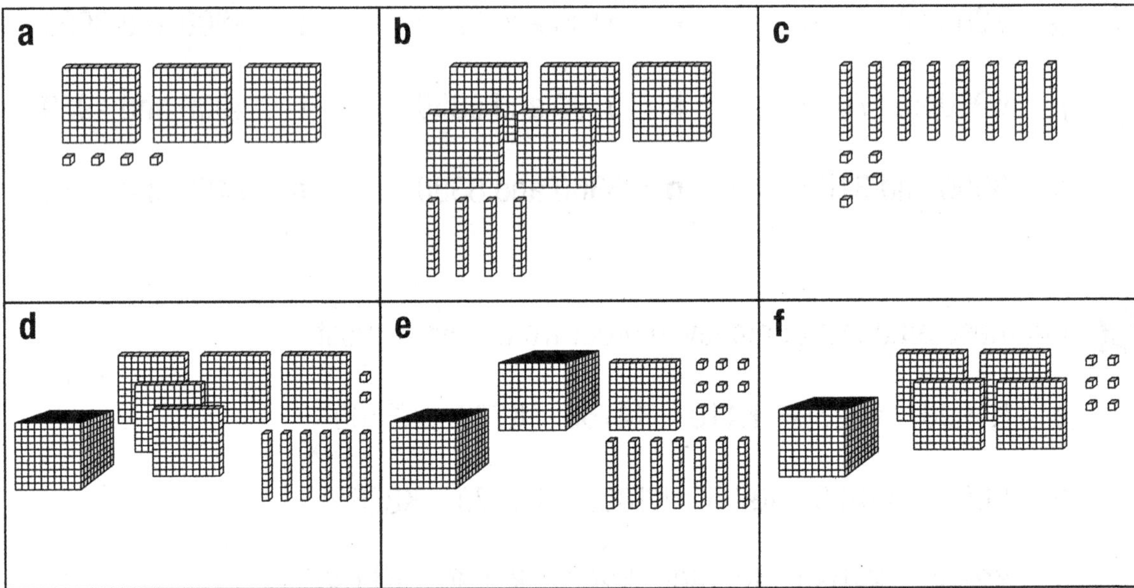

2 What number is shown on each abacus?

a
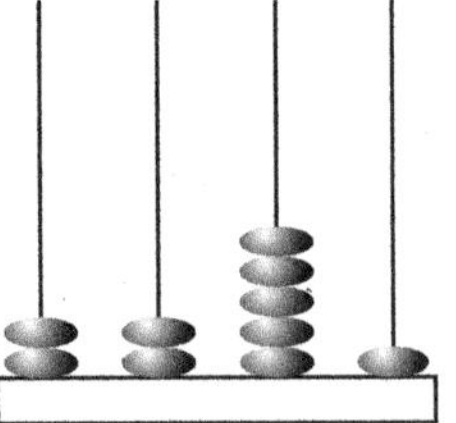

b
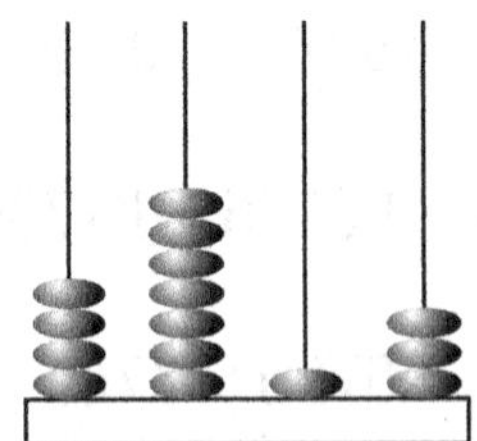

c
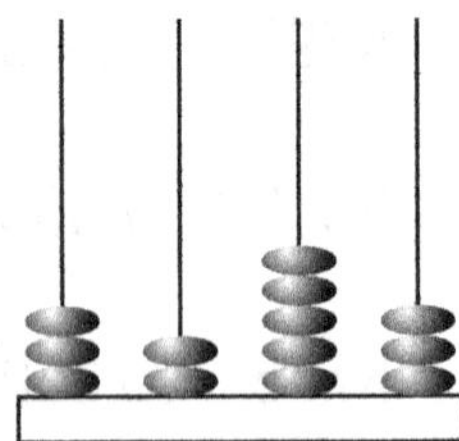

d
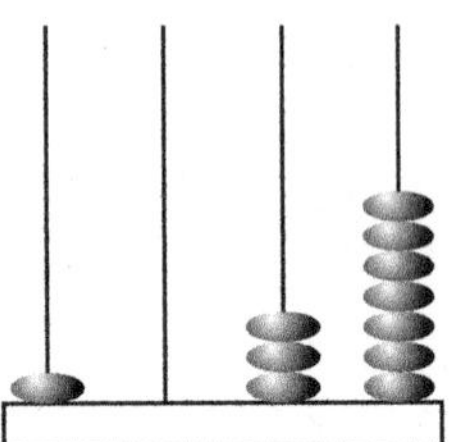

e
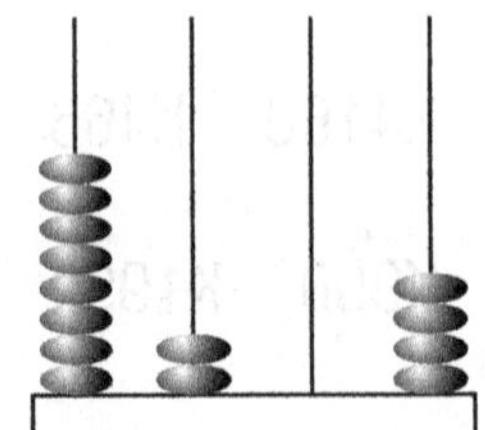

f
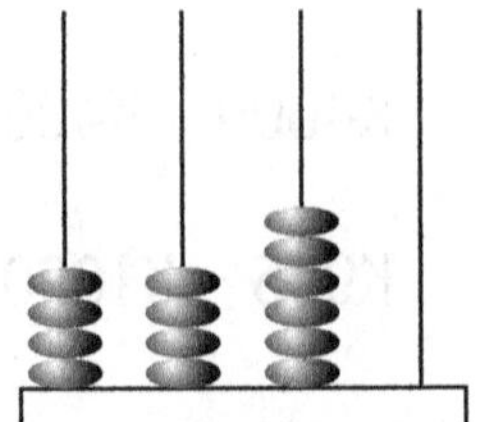

Make three and four-digit numbers

What numbers are these?

1 500 40 8

2 9 300 60

3 70 5 600

4 100 2 20

5 800 6 10

6 30 200 7

7 50 400

8 1 90 300

9 3 100 80

10 700 5

11 5 hundreds 2 thousands 3 ones 5 tens

12 2 tens 8 hundreds 7 ones 5 thousands

13 6 ones 1 thousand 3 hundreds 2 tens

14 2 thousands 5 ones 1 ten 6 hundreds

15 8 hundreds 7 thousands 4 ones

16 8 tens 3 hundreds 4 thousands

17 9 hundreds 1 one 6 thousands 5 tens

18 8 tens 5 hundreds 2 ones 1 thousand

Use digits to make numbers

1 Write the biggest and the smallest number you can make with each group of digits.

a 1 5 2

b 8 5 6

c 3 7 4

d 2 0 3

e 6 9 6

f 9 2 8

2 Write the biggest and the smallest number you can make with each group of digits.

a 7 1 6 3

b 5 8 2 7

c 9 0 4 2

d 1 4 4 8

e 6 5 0 3

f 7 2 9 5

3 Write all the numbers you can make with these digits.

a 2 9 4

b 6 6 8

4 4 9 2 3

a Write all the numbers you can make with these digits.

b Write all the numbers you can make with 9 in the thousands place.

c Write all the numbers you can make with 3 in the tens place.

Recognise the value of digits

Write the numbers that **do not fit** in each group.

1

6 hundreds
643 216 5614
5267 4612
869 1605 3868

2

5 tens
158 3250 597
5400 8054
352 1951 560

3

3 ones
5143 635 3776
4053 8312
1993 831 2803

4

4 hundreds
2457 540 1490
341 8407
4488 5412 495

5

7 thousands
755 7605 1738
7213 5764
172 7893 2076

6

9 ones
5649 139 297
1926 7859
3298 1189 698

7

2 tens
1023 652 3527
4261 8020
7825 920 5233

8

1 thousand
1789 6128 153
4614 1216
3018 188 1335

Count to 1000

1 Which flowers would the bee land on if it counted by 7s from 500?

2 Count by 8 to complete the number sequences.

a 452 ______ ______ ______ ______ ______

b 203 ______ ______ ______ ______ ______

c 597 ______ ______ ______ ______ ______

d 780 ______ ______ ______ ______ ______

e 391 ______ ______ ______ ______ ______

f 618 ______ ______ ______ ______ ______

g 926 ______ ______ ______ ______ ______

3 Count by 6 to complete the number sequences.

a 195 ______ ______ ______ ______ ______

b 898 ______ ______ ______ ______ ______

c 623 ______ ______ ______ ______ ______

d 916 ______ ______ ______ ______ ______

e 741 ______ ______ ______ ______ ______

f 267 ______ ______ ______ ______ ______

g 509 ______ ______ ______ ______ ______

Count to 10 000

1 Count by 9 to complete the number sequences.

a 2173 ______ ______ ______ ______ ______

b 8979 ______ ______ ______ ______ ______

c 1058 ______ ______ ______ ______ ______

d 5114 ______ ______ ______ ______ ______

e 9260 ______ ______ ______ ______ ______

f 3727 ______ ______ ______ ______ ______

g 7385 ______ ______ ______ ______ ______

2 Count by 7 to complete the number sequences.

a 4238 ______ ______ ______ ______ ______

b 7220 ______ ______ ______ ______ ______

c 1655 ______ ______ ______ ______ ______

d 6987 ______ ______ ______ ______ ______

e 2979 ______ ______ ______ ______ ______

f 5003 ______ ______ ______ ______ ______

g 8874 ______ ______ ______ ______ ______

Apply and use the four operations to do calculations with three and four-digit numbers

Use mental strategies for addition

1 Add each set of numbers in your head and write the answer. Look for quick ways to do the additions.

a 6 + 5 + 3 + 5 + 7 + 4

b 9 + 4 + 9 + 8 + 2 + 6

c 8 + 8 + 7 + 7 + 6 + 6

d 2 + 3 + 4 + 5 + 6 + 7

e 6 + 7 + 9 + 6 + 7 + 9

f 9 + 8 + 7 + 6 + 5 + 4

g 4 + 8 + 2 + 6 + 5 + 5

h 1 + 2 + 3 + 4 + 5 + 6

i 1 + 3 + 5 + 7 + 9 + 11

j 2 + 4 + 6 + 8 + 10 + 12

2 Copy and complete each grid. Add each pair of numbers in your head.

a

+	32	27	49	18
9				
7				
12				
20				
30				
25				

b

+	39	55	74	46
8				
6				
11				
15				
19				
29				

3 Work out the answers to these additions in your head.

a	20 + 80	**b**	25 + 25	**c**	50 + 15	**d**	20 + 75
e	45 + 50	**f**	30 + 35	**g**	33 + 20	**h**	60 + 35
i	29 + 50	**j**	36 + 40	**k**	40 + 53	**l**	75 + 25
m	56 + 50	**n**	40 + 48	**o**	27 + 40	**p**	39 + 49
q	37 + 25	**r**	63 + 23	**s**	55 + 15	**t**	29 + 16
u	28 + 51	**v**	36 + 49	**w**	18 + 69	**x**	58 + 16

4 Write three numbers in each group that total 100.

a

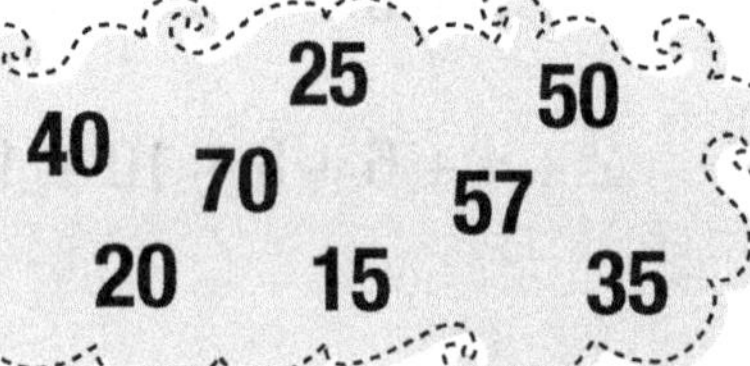

b

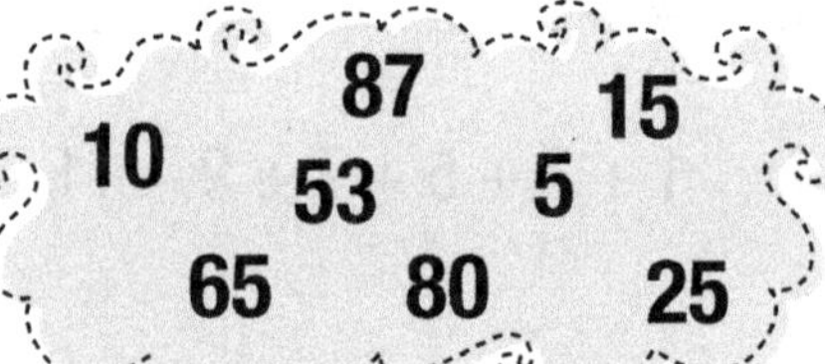

c

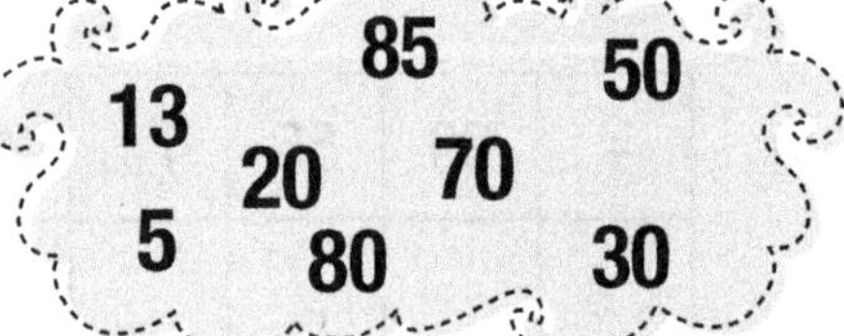

d

e

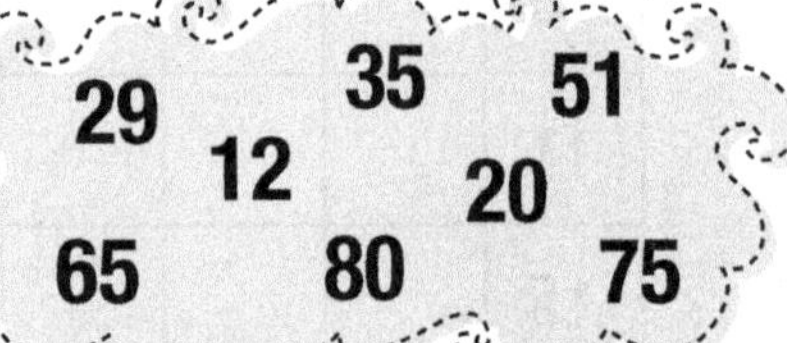

f

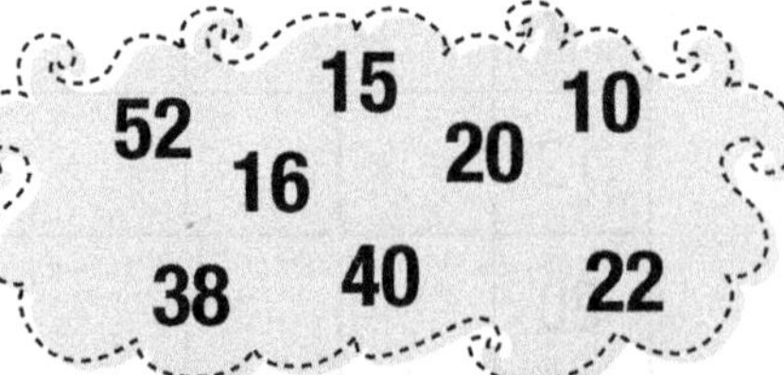

g

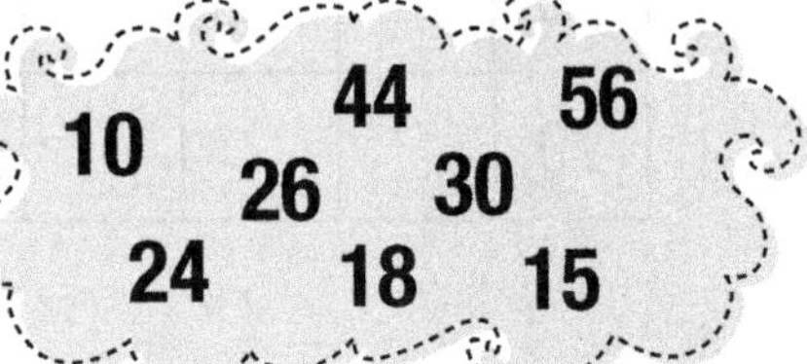

h

Use mental strategies for subtraction

1 Copy and complete these subtraction grids.

a

12 15
18 6 11
9
14 16
17 19

b

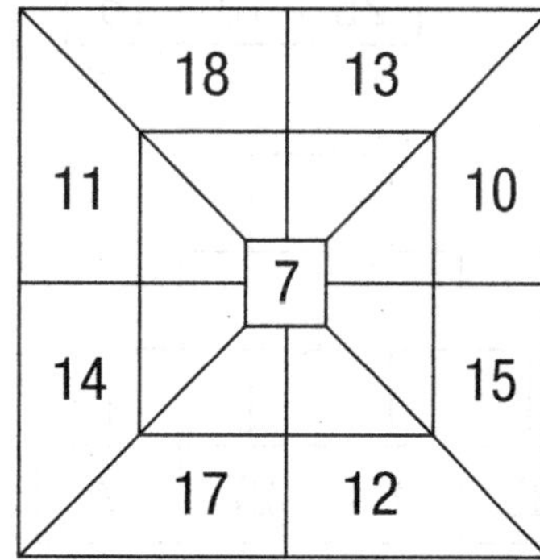

c

9 12
11 14
5
15 8
10 13

d

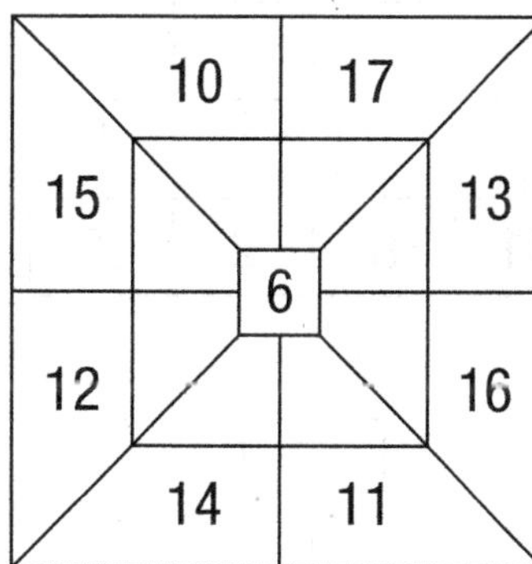

e

11 16
15 20
8
18 13
12 19

f

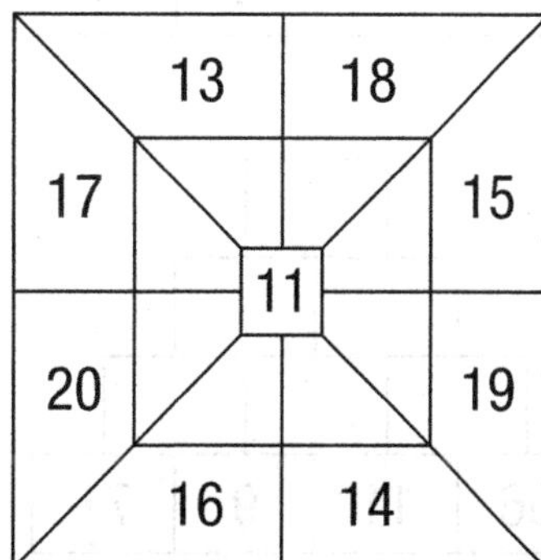

2 Copy and complete these difference pyramids by writing the missing numbers.

For example:

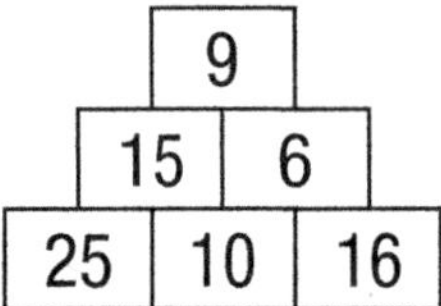

a

b

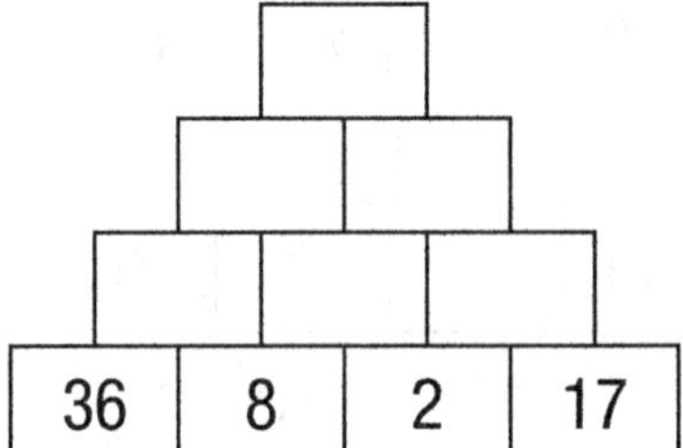

c

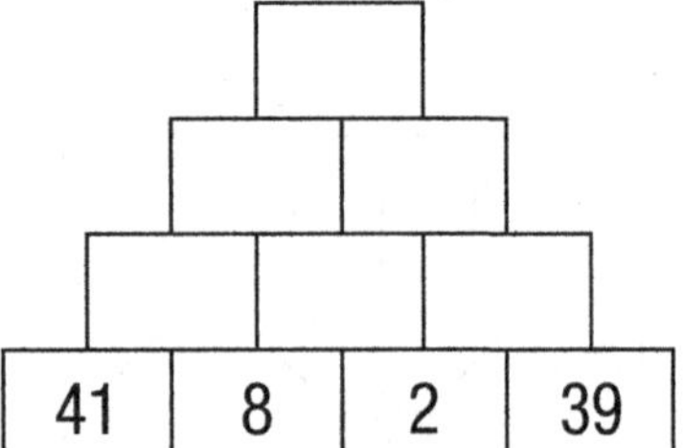

d

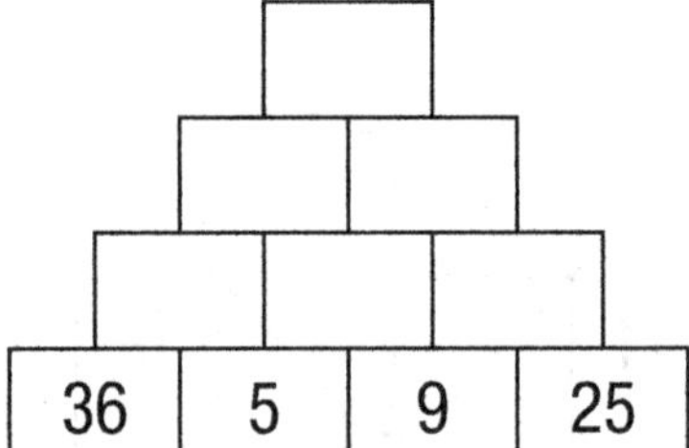

e

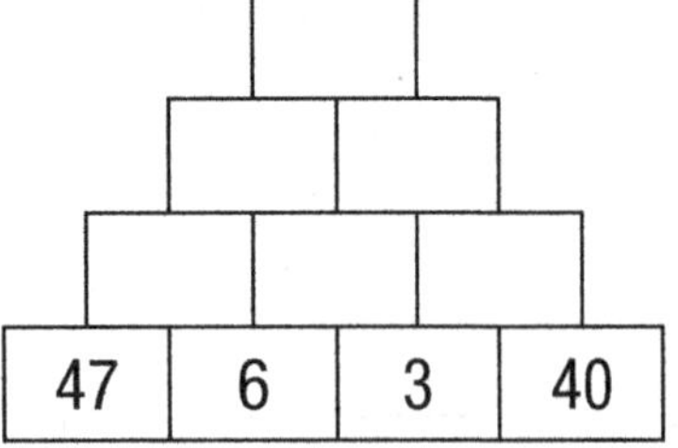

f

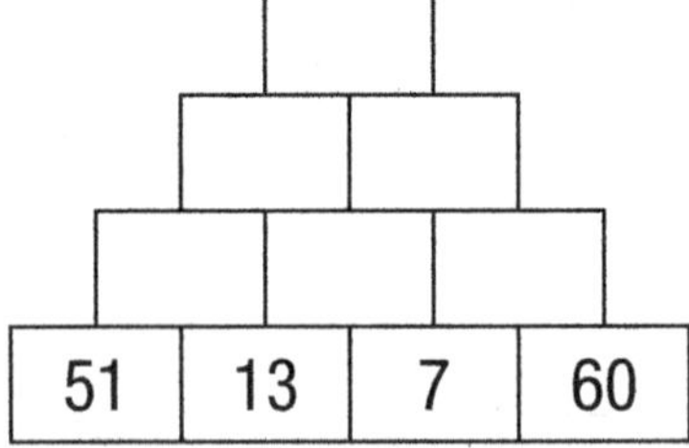

g

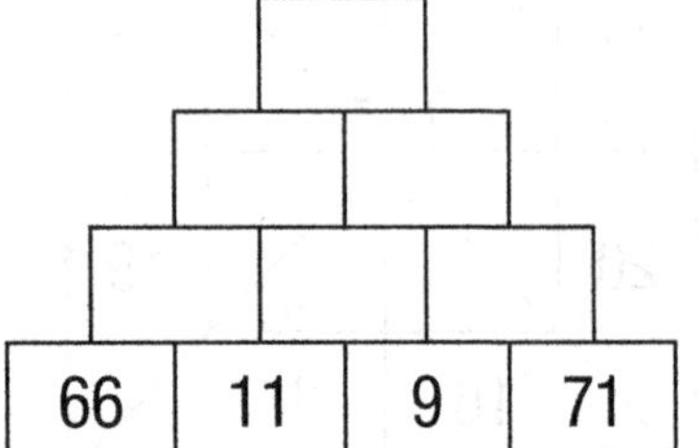

h

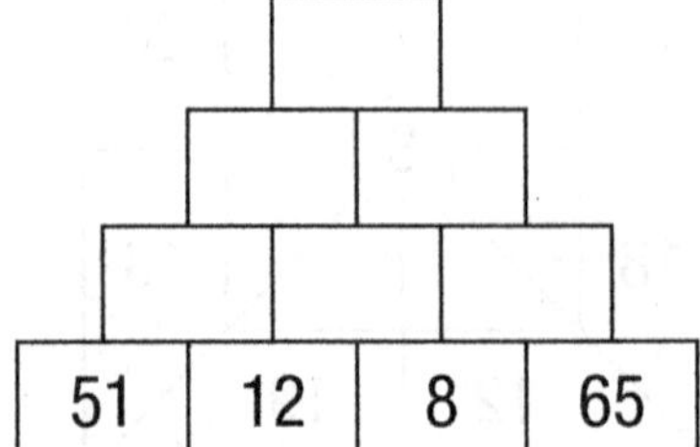

Use mental strategies for addition and subtraction

Copy and complete these diagrams.

1

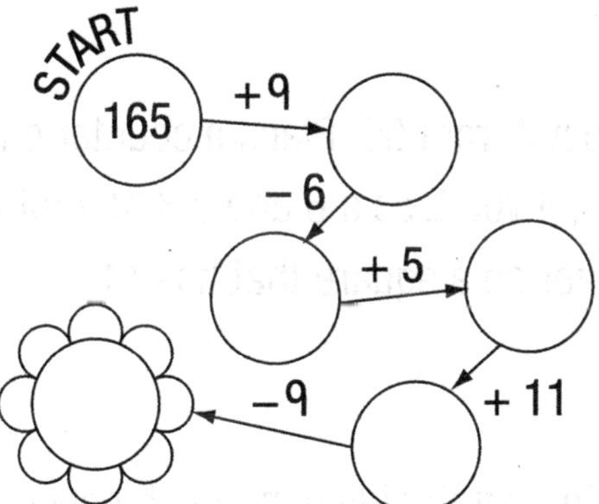

2

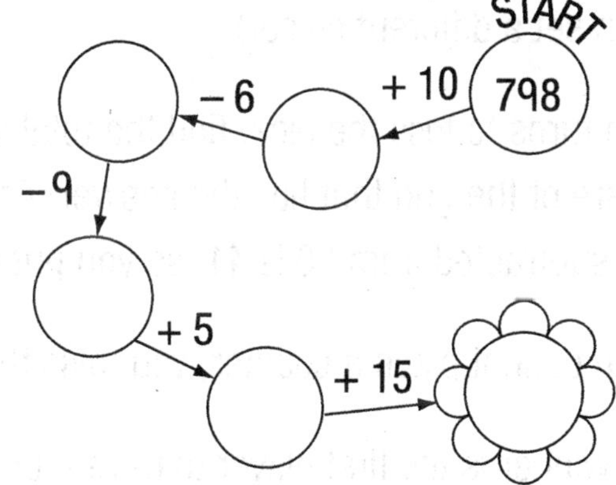

3

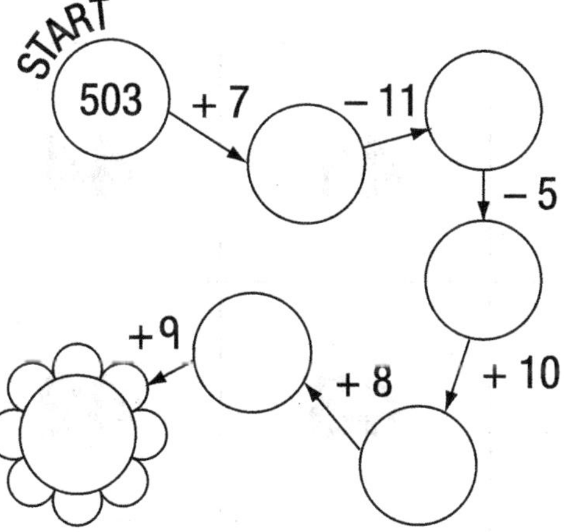

4

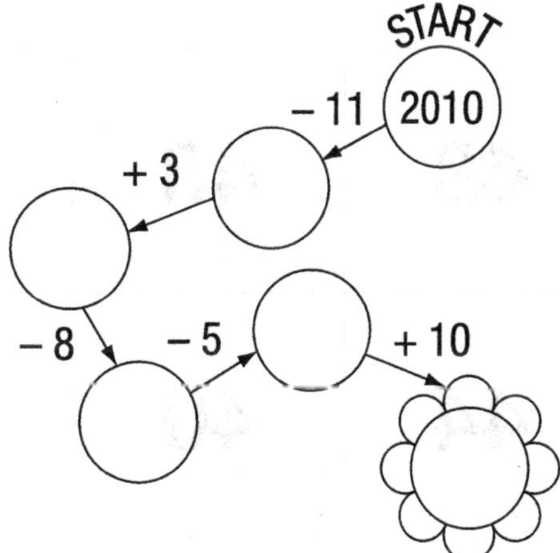

5

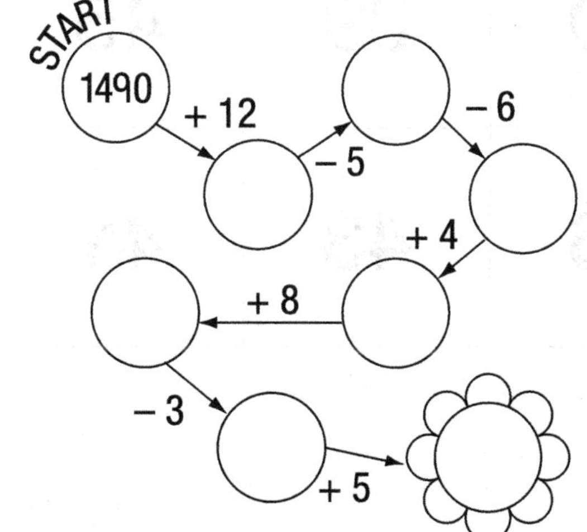

6

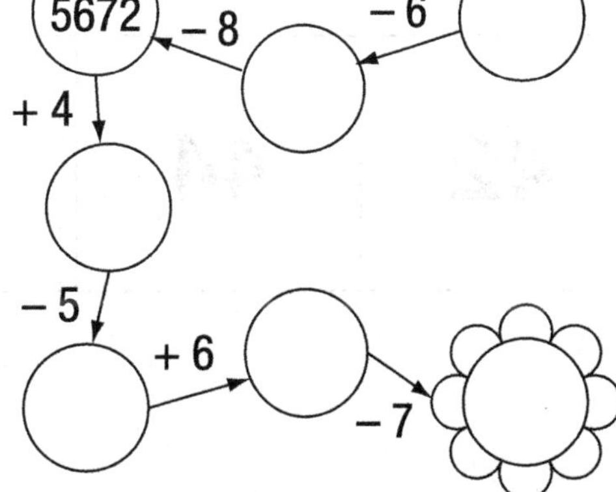

Mental strategies for addition and subtraction

Play this game with a friend. You need two dice and each player needs a set of counters (each set a different colour).

Take turns to toss the dice, find the total and subtract it from 50. Place a counter on the square of the grid that has the answer. For example, if you toss a 6 and a 3, the total is 9. 9 subtracted from 50 is 41, so you put your counter on a square that has 41.

If you cannot place a counter, you miss that turn.

The winner is the first player to have four counters in a straight line in any direction.

48	43	47	41	44
44	40	38	43	42
45	42	40	39	45
42	44	46	43	41

Use mental strategies for multiplication and division

1 Copy and complete the multiplication and division wheels.

a

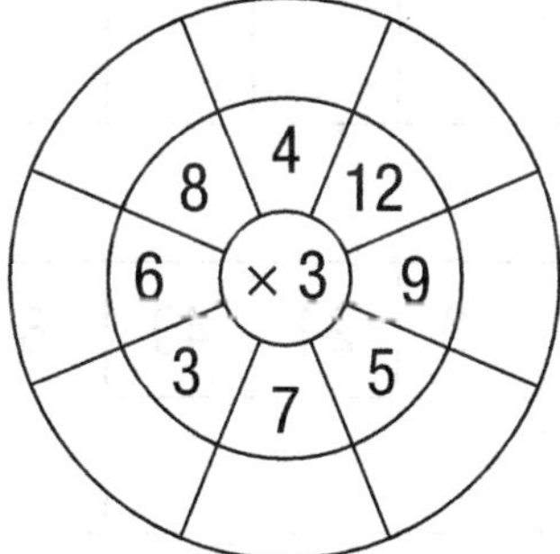

b

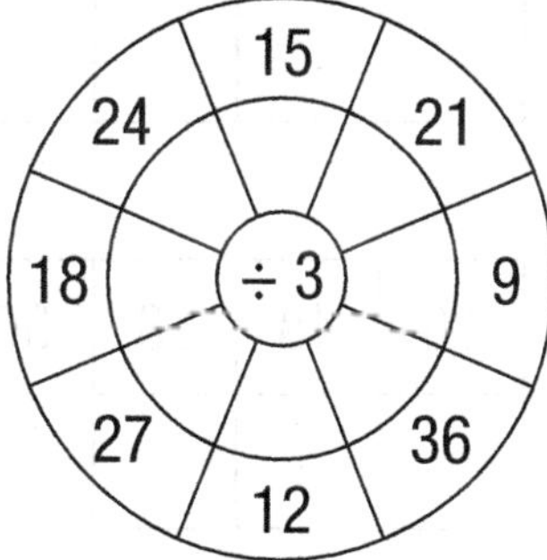

c

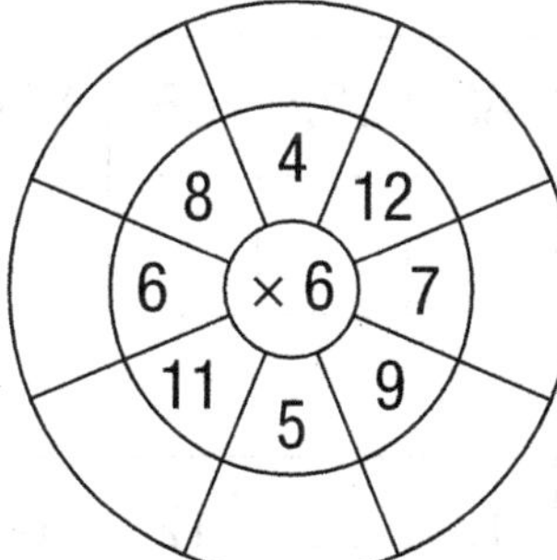

d

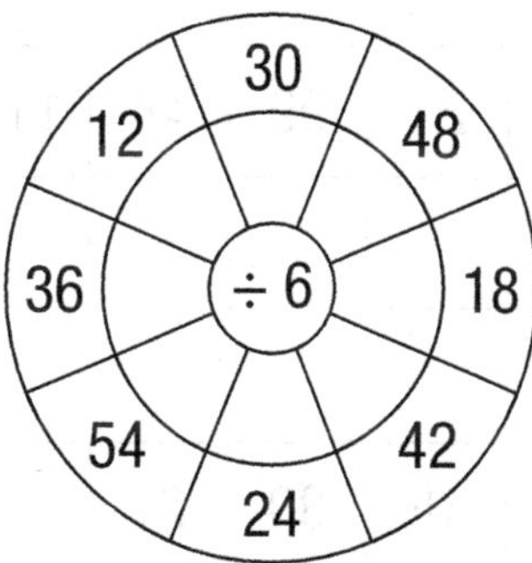

e

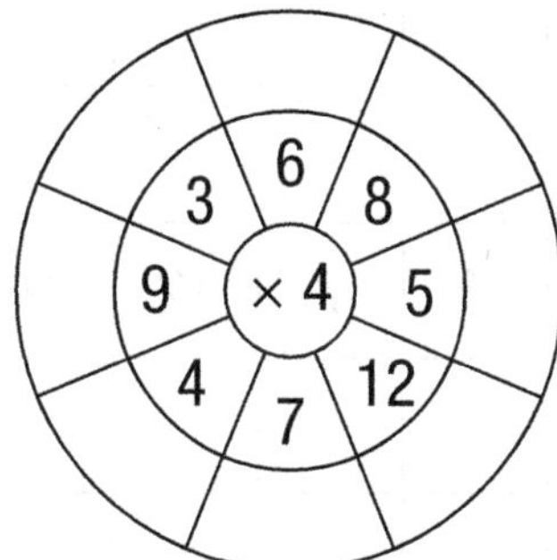

f

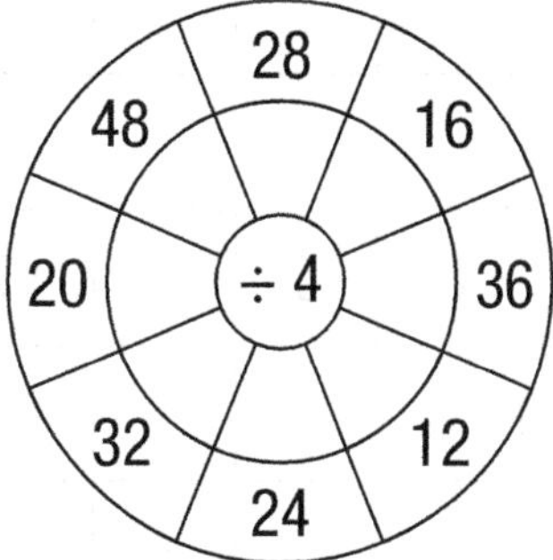

g

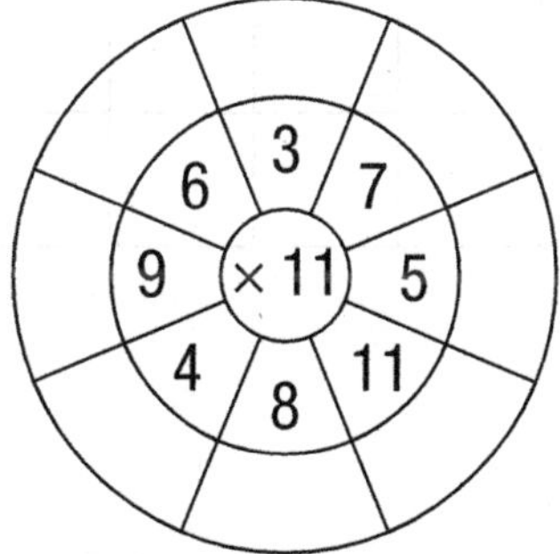

h

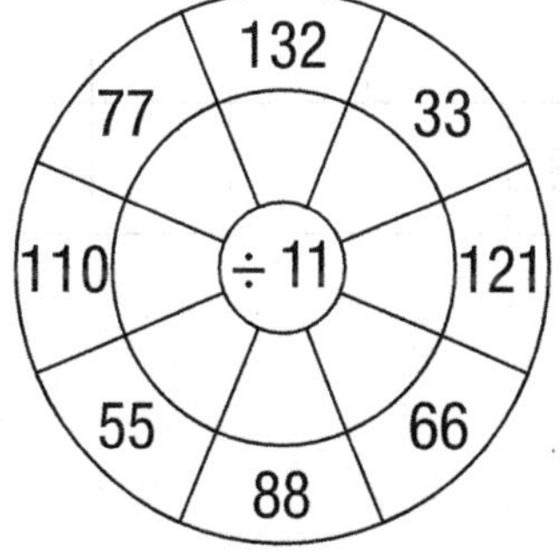

2 Copy and complete the multiplication and division grids.

a

÷	42	18	60	30	48	24	54	72	36	66	12
6											

b

÷	16	48	28	36	20	32	44	12	24	40	8
4											

c

÷	22	99	44	110	77	55	33	88	132	66	121
11											

d

÷	18	30	21	9	27	36	24	15	33	12	6
3											

e

×	8	5	2	11	6	3	10	4	7	12	9
11											
4											
6											
3											

3 Fill the gaps.

a	$4 \times 8 =$ _____	**b**	$6 \times 7 =$ _____	**c**	$24 \div 4 =$ _____
d	$36 \div 6 =$ _____	**e**	$7 \times 11 =$ _____	**f**	$27 \div 3 =$ _____
g	$3 \times 9 =$ _____	**h**	$8 \times 6 =$ _____	**i**	$88 \div 11 =$ _____
j	$7 \times 4 =$ _____	**k**	$72 \div 6 =$ _____	**l**	$54 \div 6 =$ _____
m	$4 \times 9 =$ _____	**n**	$6 \times 6 =$ _____	**o**	$32 \div 4 =$ _____
p	$48 \div 4 =$ _____	**q**	$11 \times 12 =$ _____	**r**	$3 \times 7 =$ _____
s	$30 \div 6 =$ _____	**t**	$4 \times 3 =$ _____	**u**	$110 \div 11 =$ _____
v	$6 \times 9 =$ _____				

4 Fill the gaps.

a	$6 \times$ _____ $= 42$	**b**	$16 \div$ _____ $= 4$	**c**	_____ $\times 3 = 27$
d	_____ $\div 11 = 5$	**e**	$36 \div$ _____ $= 6$	**f**	_____ $\times 6 = 72$
g	$18 \div$ _____ $= 6$	**h**	_____ $\div 4 = 9$	**i**	$11 \times$ _____ $= 121$
j	$21 \div$ _____ $= 7$	**k**	_____ $\times 8 = 48$	**l**	_____ $\times 11 = 110$
m	$15 \div$ _____ $= 5$	**n**	$6 \times$ _____ $= 24$	**o**	_____ $\times 3 = 33$
p	_____ $\div 6 = 9$	**q**	$8 \times$ _____ $= 32$	**r**	$132 \div$ _____ $= 11$
s	_____ $\div 4 = 5$	**t**	_____ $\times 6 = 72$	**u**	$42 \div$ _____ $= 7$
v	_____ $\times 3 = 36$				

5 Copy each number chain and show how to move from one number to the next by multiplying or dividing. The first one is done for you.

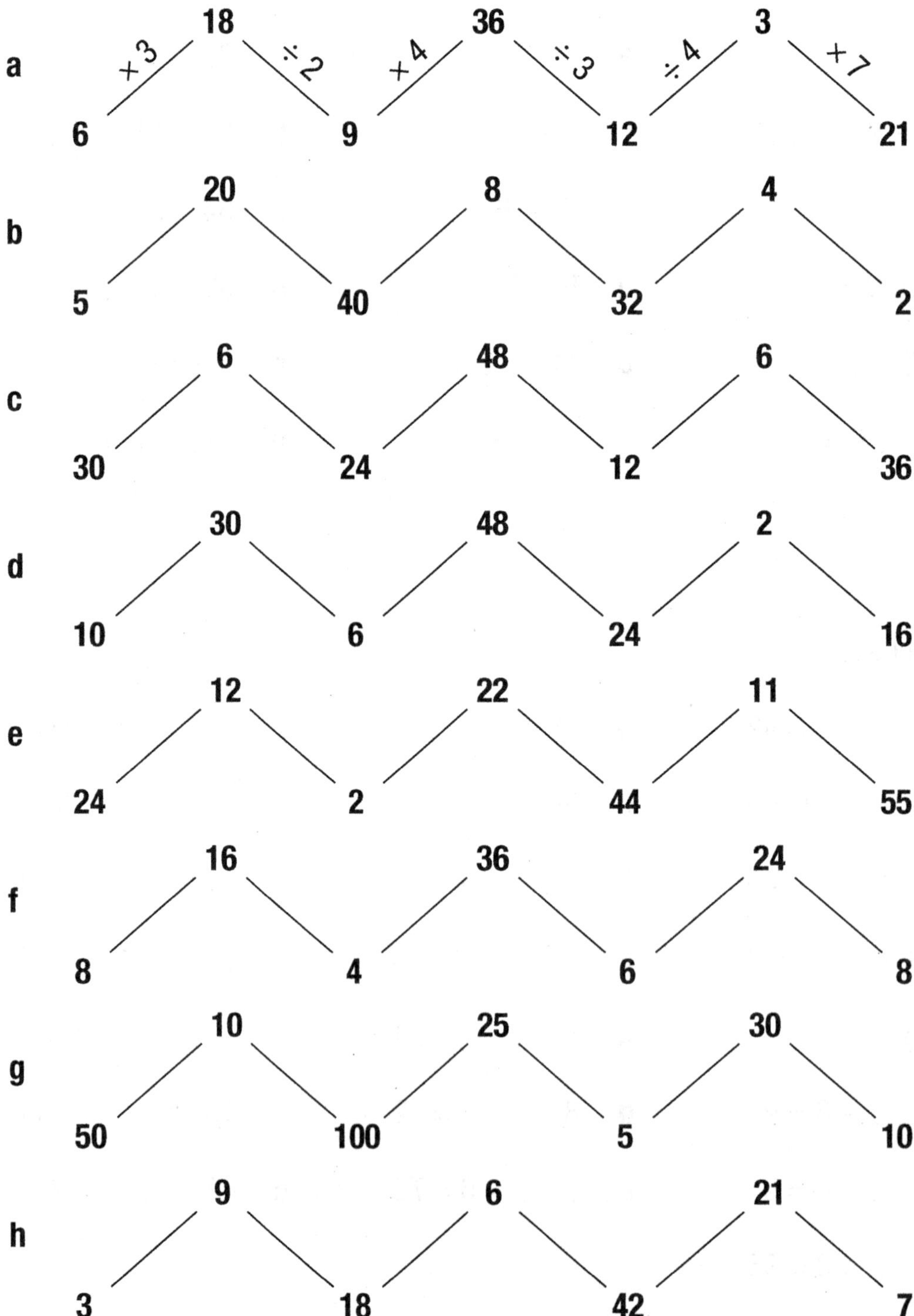

Use mental strategies for division

Play this game with a friend. Each player needs a set of about 12 counters (each set a different colour).

Choose a number from each box below the grid. Take turns to divide the first number by the second number to make one of the answers on the grid. Cover that answer with one of your counters.

The winner is the first player to get four counters in a row in any direction. Only one player can have a counter on any one answer.

3	4	10	12	5	10
2	4	8	6	2	7
5	9	20	1	3	5
50	2	3	7	6	2
10	6	3	1	9	20

First box		Second box
15 100		5 7
36 55		10 2 4
10 35 18		8 9
24 12		3
88 21		11 6

Multiply two-digit numbers by 10

Copy and complete each multiplication.

1

a $8 \times 10 =$ ____ b $12 \times 10 =$ ____ c $25 \times 10 =$ ____

d $19 \times 10 =$ ____ e $34 \times 10 =$ ____ f $61 \times 10 =$ ____

g $15 \times 10 =$ ____ h $47 \times 10 =$ ____ i $50 \times 10 =$ ____

j $78 \times 10 =$ ____ k $23 \times 10 =$ ____ l $86 \times 10 =$ ____

m $52 \times 10 =$ ____ n $37 \times 10 =$ ____ o $95 \times 10 =$ ____

p $28 \times 10 =$ ____ q $49 \times 10 =$ ____ r $80 \times 10 =$ ____

s $55 \times 10 =$ ____ t $91 \times 10 =$ ____ u $33 \times 10 =$ ____

v $67 \times 10 =$ ____ w $42 \times 10 =$ ____

2

a $38 \times 10 =$ ____ b ____ $\times 10 = 560$ c $97 \times 10 =$ ____

d ____ $\times 10 = 710$ e ____ $\times 10 = 630$ f $45 \times 10 =$ ____

g ____ $\times 10 = 790$ h $21 \times 10 =$ ____ i $88 \times 10 =$ ____

j ____ $\times 10 = 300$ k ____ $\times 10 = 140$ l $59 \times 10 =$ ____

m $234 \times 10 =$ ____ n $156 \times 10 =$ ____ o $547 \times 10 =$ ____

p $311 \times 10 =$ ____ q $408 \times 10 =$ ____ r ____ $\times 10 = 2300$

s ____ $\times 10 = 1480$ t ____ $\times 10 = 1040$ u $710 \times 10 =$ ____

v $118 \times 10 =$ ____ w ____ $\times 10 = 6000$

List multiples of a given number

1 Write the numbers that are multiples of each number on the left.

a	**4**	16	22	35	12	20	28	31	36	40	25
b	**6**	25	18	36	15	27	42	60	13	33	66
c	**3**	36	14	25	33	16	24	9	21	15	38
d	**5**	35	23	30	36	45	27	14	50	55	17
e	**7**	15	49	63	56	35	21	29	30	14	70
f	**11**	21	55	88	37	22	51	17	33	40	66
g	**10**	25	80	60	47	15	90	100	54	30	20
h	**8**	46	16	28	33	40	21	64	56	24	88
i	**9**	99	27	36	16	90	72	56	81	63	47
j	**12**	24	10	48	60	17	36	30	20	120	18

2 Write the number that is **not** a multiple of the number on the left.

a	**11**	55	88	110	22	99	121	33	132	111
b	**5**	40	25	100	55	30	17	45	120	150
c	**2**	28	14	50	32	18	100	20	45	80
d	**6**	18	28	42	60	36	12	54	30	24
e	**4**	24	44	32	42	16	40	100	12	28
f	**10**	50	90	20	125	30	80	200	160	320
g	**3**	15	30	19	45	21	33	300	27	90
h	**8**	24	40	18	88	32	48	80	16	120

Add three-digit numbers

Help Box

Add whole numbers without trading:

352 + 433

$$\begin{array}{r} 352 \\ +\ 433 \\ \hline 785 \end{array}$$

Add whole numbers with trading in one or more places:

287 + 576

$$\begin{array}{r} 287 \\ +\ {}_{1}5{}_{1}76 \\ \hline 863 \end{array}$$

Set out and solve these additions to see how many pineapples you can pick.

1 $\begin{array}{r} 524 \\ +\ 273 \\ \hline \end{array}$

2 $\begin{array}{r} 192 \\ +\ 603 \\ \hline \end{array}$

3 $\begin{array}{r} 255 \\ +\ 442 \\ \hline \end{array}$

4 $\begin{array}{r} 720 \\ +\ 178 \\ \hline \end{array}$

5 $\begin{array}{r} 316 \\ +\ 581 \\ \hline \end{array}$

6 $\begin{array}{r} 427 \\ +\ 362 \\ \hline \end{array}$

7 $\begin{array}{r} 641 \\ +\ 136 \\ \hline \end{array}$

8 $\begin{array}{r} 552 \\ +\ 427 \\ \hline \end{array}$

9 $\begin{array}{r} 358 \\ +\ 275 \\ \hline \end{array}$

10 $\begin{array}{r} 215 \\ +\ 567 \\ \hline \end{array}$

11 $\begin{array}{r} 567 \\ +\ 328 \\ \hline \end{array}$

12 $\begin{array}{r} 525 \\ +\ 295 \\ \hline \end{array}$

13 $\begin{array}{r} 738 \\ +\ 516 \\ \hline \end{array}$

14 $\begin{array}{r} 580 \\ +\ 637 \\ \hline \end{array}$

15 $\begin{array}{r} 456 \\ +\ 366 \\ \hline \end{array}$

16 $\begin{array}{r} 573 \\ +\ 637 \\ \hline \end{array}$

How many pineapples did you pick? _____

Add three and four-digit numbers

Set out and solve these additions to see how many fish you can spear.

1

2

3

4

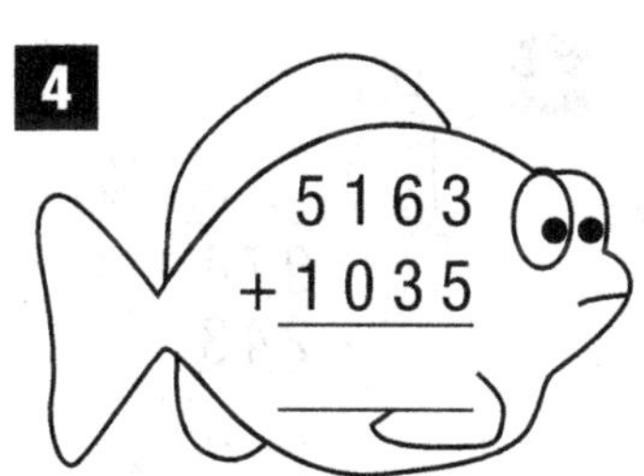

5

6

7

8

9

10

11

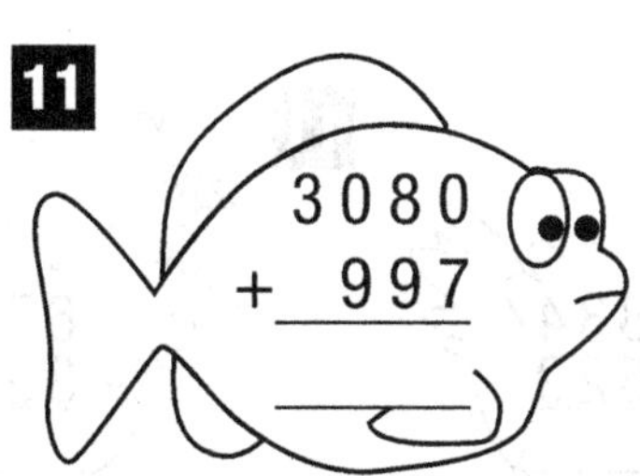

12

How many fish did you spear? _____

Subtract three-digit numbers

Help Box

Subtract whole numbers without trading: 769 – 524

```
  7 6 9
- 5 2 4
  2 4 5
```

Subtract whole numbers with trading in one or more places: 825 – 358

```
  7 11 1
  8̸ 2̸ 5
- 3  5  8
  4  6  7
```

Set out and solve these subtractions to see how many chickens you can catch.

1

2

3

4

5

6

7

8

9

10

11

12

How many chickens did you catch? _____

Subtract three and four-digit numbers

Set out and solve these subtractions to see how many targets you can hit.

1

2
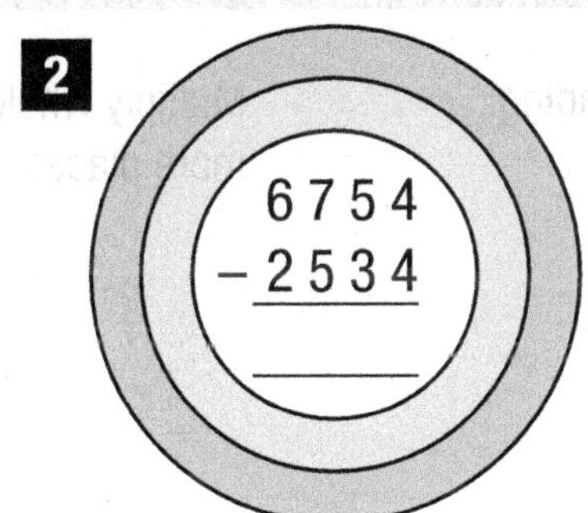

3
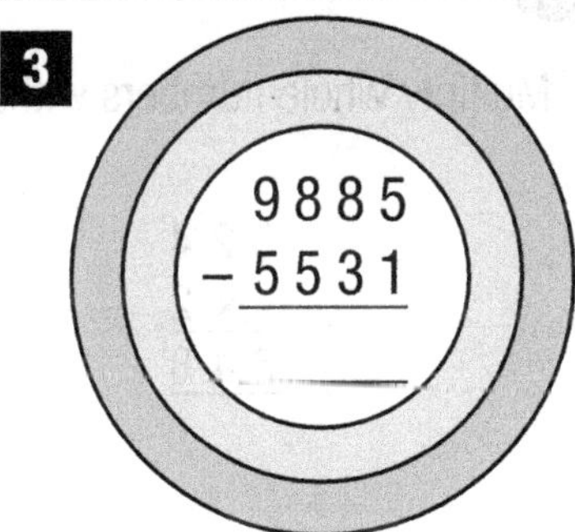

4
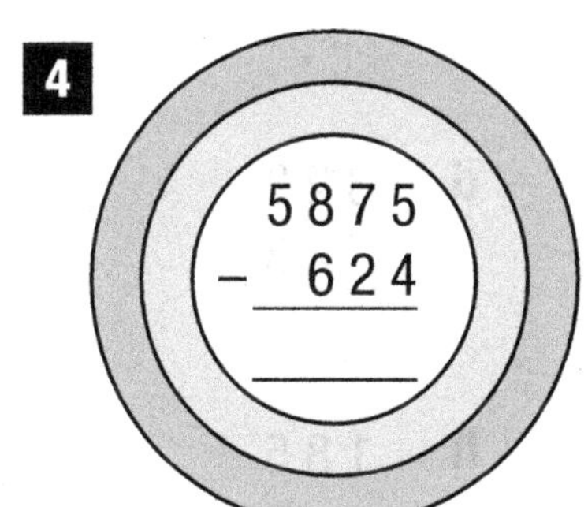

5

6
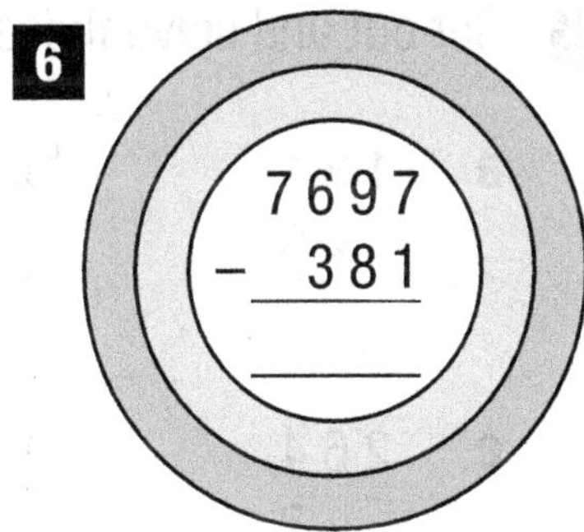

7
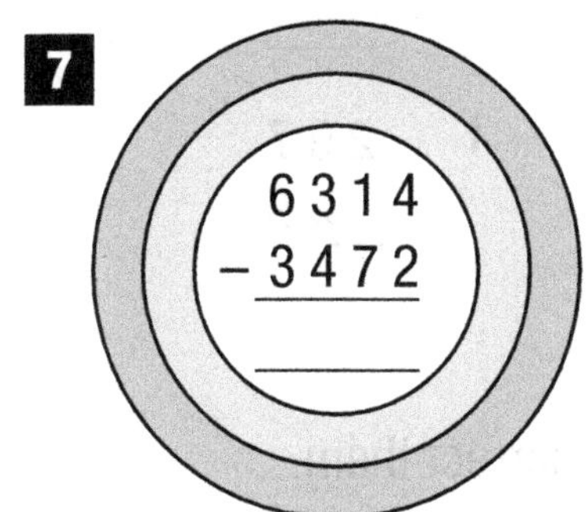

8

9
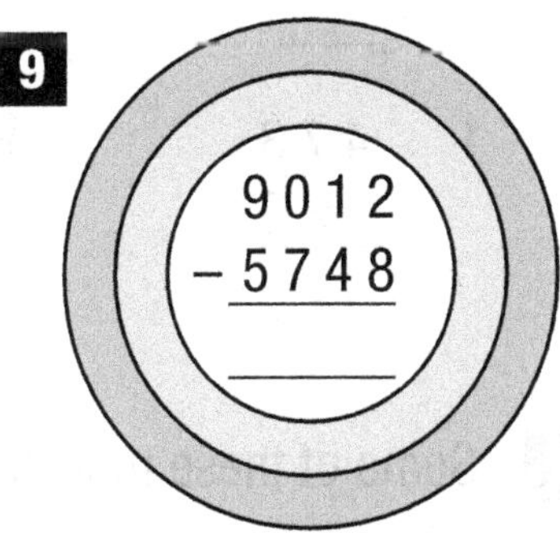

10
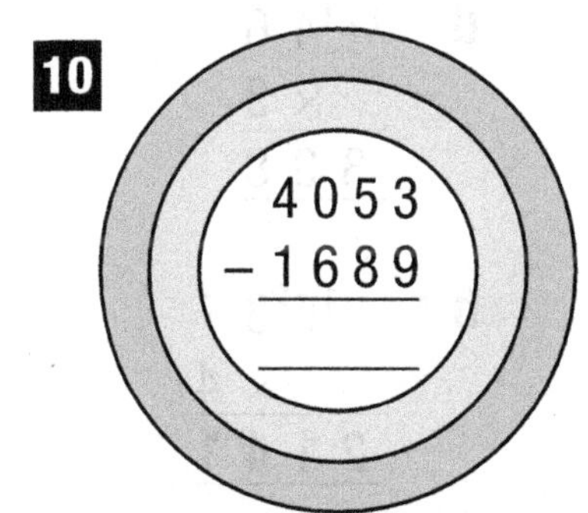

11
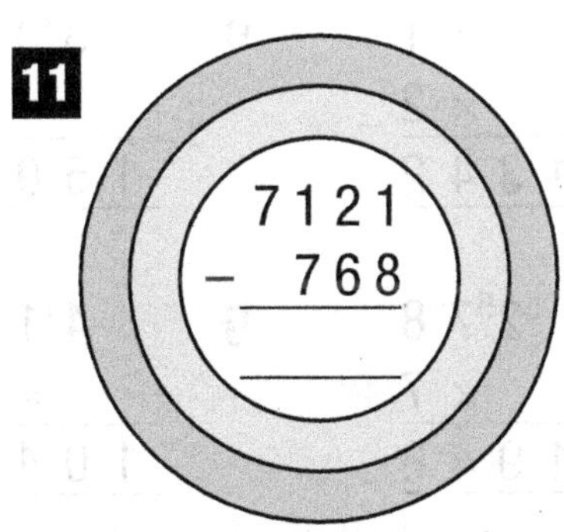

12
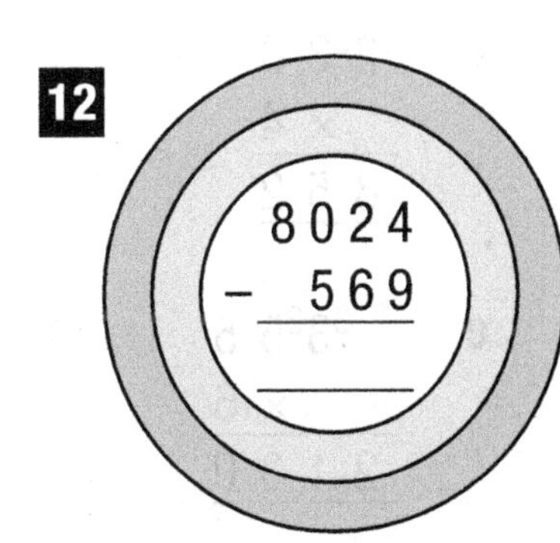

Multiply three-digit numbers

Help Box

Multiply whole numbers without trading:

$$\begin{array}{r} 324 \\ \times\ 2 \\ \hline 648 \end{array}$$

Multiply whole numbers with trading in one or more places:

$$\begin{array}{r} {}^{3}2\,{}^{3}6\,7 \\ \times\ 5 \\ \hline 1335 \end{array}$$

1 Set out and solve these multiplications.

a $\begin{array}{r} 131 \\ \times 3 \\ \hline \end{array}$ **b** $\begin{array}{r} 244 \\ \times 2 \\ \hline \end{array}$ **c** $\begin{array}{r} 413 \\ \times 2 \\ \hline \end{array}$ **d** $\begin{array}{r} 323 \\ \times 3 \\ \hline \end{array}$

e $\begin{array}{r} 264 \\ \times 2 \\ \hline \end{array}$ **f** $\begin{array}{r} 318 \\ \times 3 \\ \hline \end{array}$ **g** $\begin{array}{r} 286 \\ \times 4 \\ \hline \end{array}$ **h** $\begin{array}{r} 185 \\ \times 5 \\ \hline \end{array}$

i $\begin{array}{r} 473 \\ \times 6 \\ \hline \end{array}$ **j** $\begin{array}{r} 394 \\ \times 7 \\ \hline \end{array}$ **k** $\begin{array}{r} 538 \\ \times 5 \\ \hline \end{array}$ **l** $\begin{array}{r} 297 \\ \times 8 \\ \hline \end{array}$

2 Some of these multiplications are incorrect. Find them and correct them.

a $\begin{array}{r} {}^{1}2\,{}^{3}3\,8 \\ \times 4 \\ \hline 852 \end{array}$ **b** $\begin{array}{r} {}^{1}6\,7\,1 \\ \times 2 \\ \hline 1342 \end{array}$ **c** $\begin{array}{r} 3\,{}^{4}0\,9 \\ \times 5 \\ \hline 1505 \end{array}$ **d** $\begin{array}{r} {}^{1}1\,{}^{1}4\,6 \\ \times 3 \\ \hline 338 \end{array}$

e $\begin{array}{r} {}^{3}3\,{}^{3}5\,5 \\ \times 6 \\ \hline 2130 \end{array}$ **f** $\begin{array}{r} {}^{5}2\,{}^{5}7\,8 \\ \times 7 \\ \hline 1946 \end{array}$ **g** $\begin{array}{r} 4\,{}^{1}1\,5 \\ \times 3 \\ \hline 1045 \end{array}$ **h** $\begin{array}{r} {}^{1}5\,{}^{2}3\,7 \\ \times 4 \\ \hline 2141 \end{array}$

Multiply four-digit numbers

The answers to these multiplications are written below the grid. Work out each one and write the letter that matches each answer to discover the secret words.

1 R 2314 × 3 ____	**2** L 3352 × 5 ____	**3** E 1647 × 4 ____
4 O 4510 × 2 ____	**5** V 2065 × 3 ____	**6** E 3443 × 7 ____
7 Y 1804 × 6 ____	**8** D 5232 × 4 ____	**9** E 4573 × 3 ____
10 L 2868 × 5 ____	**11** W 3169 × 2 ____	**12** N 1958 × 8 ____

6195 13 719 6942 10 824 ____ ____ ____ ____

6338 24 101 14 340 16 760 ____ ____ ____ ____

20 928 9020 15 664 6588 ____ ____ ____ ____

Divide three-digit numbers

Help Box

Divide whole numbers without trading:

$3\overline{)693}$ = 231 or $3\overline{)693}$ with 231 written above

Divide whole numbers with trading in one or more places:

$5\overline{)7^{2}2^{2}5}$ = 145 or $5\overline{)7^{2}2^{2}5}$ with 145 written above

Remember

The division symbol can be written as ÷, $\overline{)\quad}$ or $\overline{)\quad}$, but each sign means the same thing.

1 Set out and solve these divisions.

a 575 ÷ 5 **b** 864 ÷ 4 **c** 675 ÷ 3

d 714 ÷ 6 **e** 432 ÷ 3 **f** 740 ÷ 5

g 826 ÷ 7 **h** 536 ÷ 4 **i** 918 ÷ 6

j $3\overline{)582}$ **k** $5\overline{)835}$ **l** $7\overline{)749}$

m $4\overline{)652}$ **n** $8\overline{)848}$ **o** $6\overline{)852}$

p $7\overline{)812}$ **q** $3\overline{)774}$ **r** $4\overline{)976}$

Remember

When the answer to a division problem has a remainder, it is written as remainder or rem.
Example: 16 ÷ 5 = 3 remainder 1 or 3 rem. 1

2 Set out and solve these divisions with remainders.

a	217 ÷ 3	**b**	634 ÷ 5	**c**	338 ÷ 6
d	527 ÷ 4	**e**	459 ÷ 7	**f**	793 ÷ 3
g	659 ÷ 8	**h**	351 ÷ 6	**i**	874 ÷ 5
j	$7\overline{)863}$	**k**	$4\overline{)467}$	**l**	$8\overline{)945}$
m	$3\overline{)544}$	**n**	$9\overline{)839}$	**o**	$6\overline{)758}$
p	$4\overline{)615}$	**q**	$5\overline{)982}$	**r**	$3\overline{)622}$
s	$6\overline{)583}$	**t**	$8\overline{)361}$	**u**	$4\overline{)277}$
v	$7\overline{)815}$	**w**	$3\overline{)764}$	**x**	$5\overline{)409}$

Divide three and four-digit numbers

Copy and complete this division grid. Colour all the boxes with an answer less than 150. What do you notice?

1 $7\overline{)265}$	**2** $9\overline{)307}$	**3** $5\overline{)1623}$	**4** $8\overline{)1165}$	**5** $6\overline{)142}$
6 $6\overline{)246}$	**7** $6\overline{)1412}$	**8** $9\overline{)2305}$	**9** $5\overline{)953}$	**10** $3\overline{)413}$
11 $9\overline{)3055}$	**12** $6\overline{)2547}$	**13** $5\overline{)638}$	**14** $8\overline{)1845}$	**15** $4\overline{)2133}$
16 $7\overline{)947}$	**17** $3\overline{)892}$	**18** $7\overline{)1065}$	**19** $6\overline{)1489}$	**20** $9\overline{)1240}$
21 $4\overline{)521}$	**22** $8\overline{)1142}$	**23** $4\overline{)2551}$	**24** $7\overline{)1016}$	**25** $5\overline{)633}$

Solve word problems

Decide what process to use and then solve each word problem.

1 One school has 197 students and another school has 236 students. What is the total number of students at both schools?

2 Josephine had K1754 in her bank account. She spent K688 on an airfare. How much money is left in her account?

3 Some tourists have travelled 216 kilometres of the 652 kilometres to reach their destination. How much further do they have to travel?

4 A PMV travels 1278 kilometres per week. How far does it travel in:

a 3 weeks? **b** 4 weeks? **c** 6 weeks? **d** 8 weeks?

5 If K955 is shared equally among five people, how much does each person get?

6 A water tank contained 1650 litres of water and 872 litres of this were used to water the garden. How much water was still in the tank?

7 Joshua plans to save K265 each month. How much would he save after:

a 4 months? **b** 5 months? **c** 7 months? **d** 9 months?

8 A truck started off carrying a load weighing 1254 kilograms. Along the way two loads were added to this, one of 876 kilograms and another of 563 kilograms. What was the total weight of the load being carried by the truck?

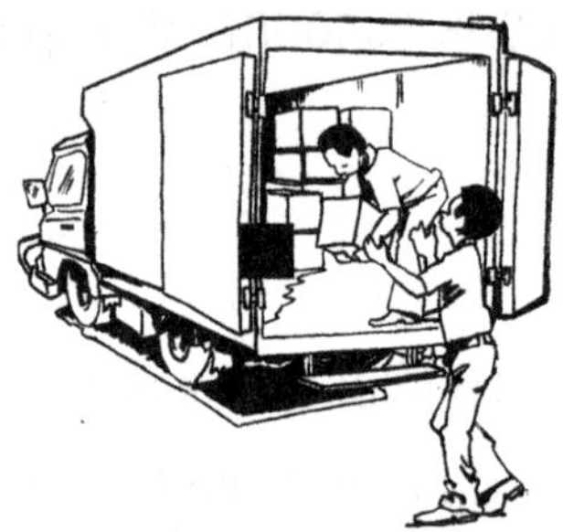

9 A hotel room for three people cost K585 per night. If the people shared the cost equally, how much would each person pay?

10 Seven tourists each spent K287 on a fishing trip. How much did they spend in total?

11 To reach our destination we had to take two flights. The first flight was a distance of 1346 kilometres and the second flight was 886 kilometres.

a How many kilometres did we travel altogether?

b What was the difference in kilometres between the first and second flight?

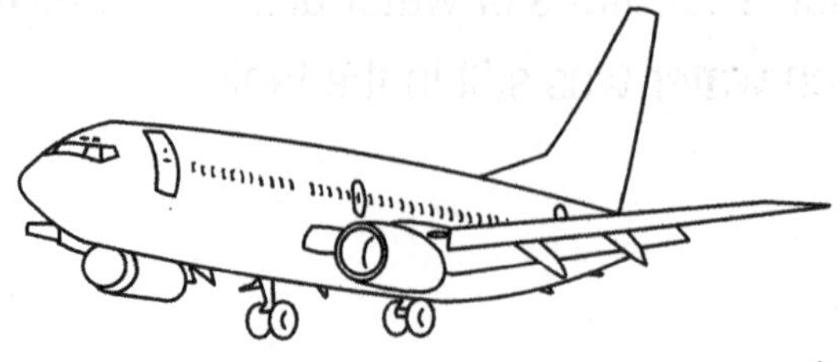

12 A hotel room costs K344 per night. How much would it cost to stay:

a 3 nights? **b** 4 nights? **c** 6 nights? **d** 7 nights?

13 The total distance travelled by a car over seven days was 2478 kilometres. The same distance was travelled each day. How far did the car travel each day?

14 In 2006, 3747 people walked the Kokoda Track. In 2007, 5146 people walked it.

a How many more walked the Kokoda Track in 2007 than 2006?

b What was the total number of people walking the Kokoda Track in 2006 and 2007?

15 A tour operator received a total of K1480 for a two-day tour. There were eight people on the tour and each paid the same amount. How much did each person pay?

16 If 441 plants were planted so there were seven in each row, how many rows would there be?

Compare and represent common fractions

Represent fractions of shapes and collections

What fraction of each shape or collection has been shaded? Write the fractions in symbols and words.

1
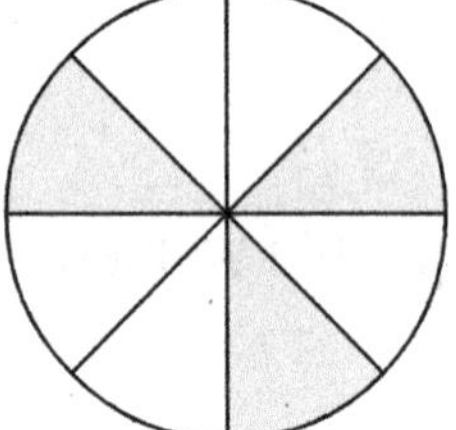

2
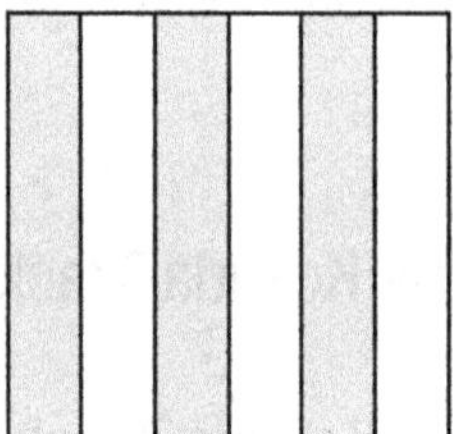

3
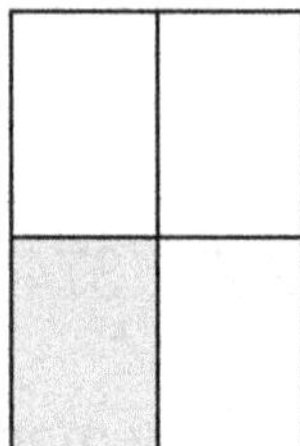

4
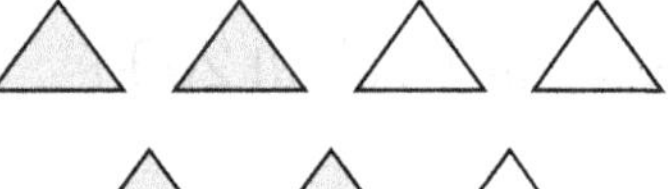

5
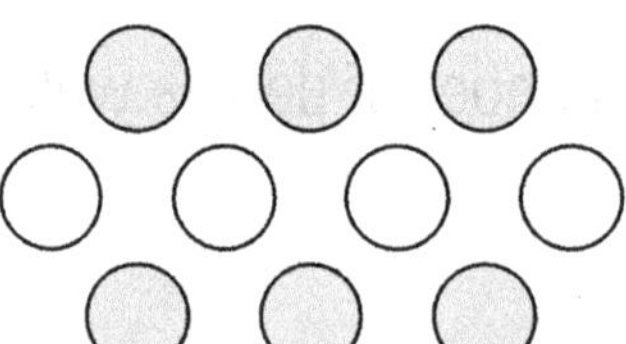

6
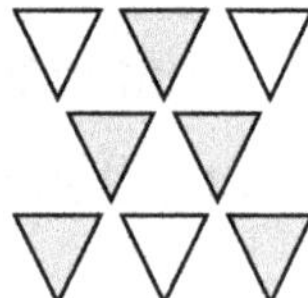

7
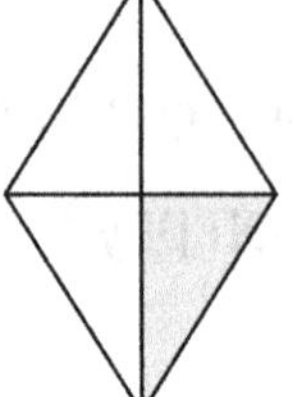

8
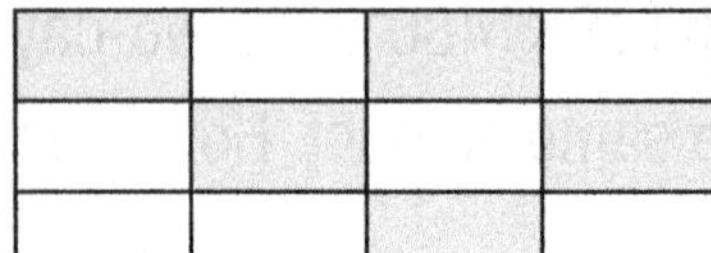

9
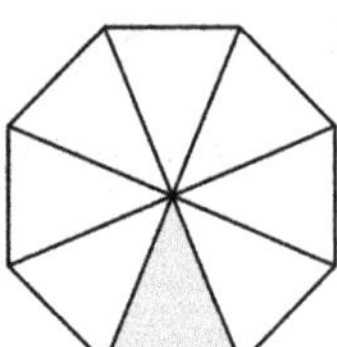

10

11
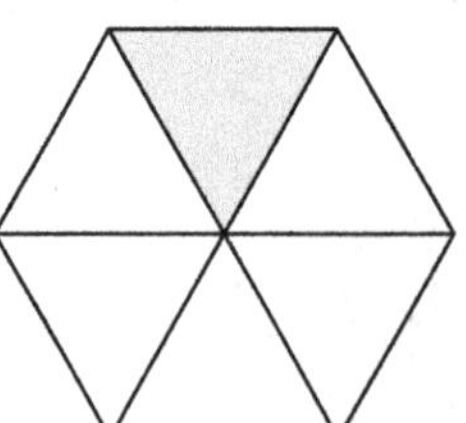

12
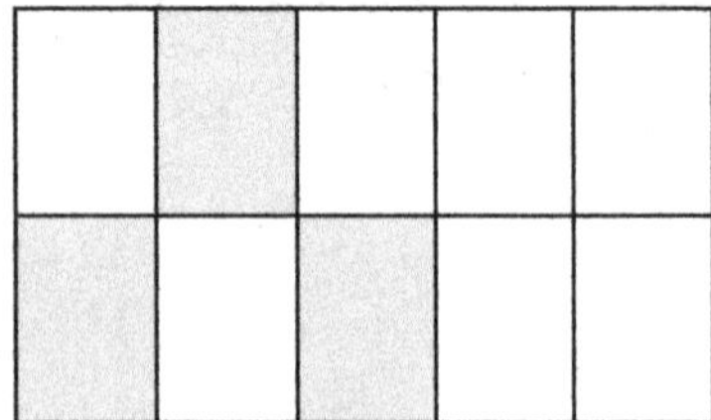

13
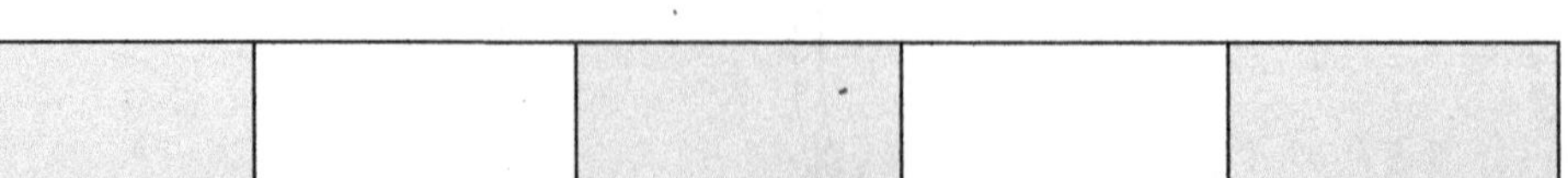

14

15
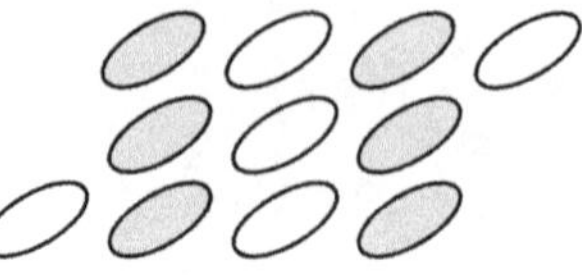

16

Write fraction number sentences

1 If each strip is worth 1, what fraction do you need to complete each strip? Write your answer as a fraction number sentence like the first example.

a

$\frac{1}{2}$	$\frac{1}{4}$	

$$\frac{1}{2} + \frac{1}{4} + \frac{1}{4} = 1$$

b

$\frac{1}{4}$	$\frac{1}{4}$	

c

$\frac{1}{4}$	$\frac{1}{8}$	$\frac{1}{8}$	$\frac{1}{4}$	$\frac{1}{8}$	

d

$\frac{1}{4}$	$\frac{1}{4}$	$\frac{1}{8}$	$\frac{1}{8}$	

e

$\frac{1}{2}$	$\frac{1}{8}$	$\frac{1}{8}$	

f

$\frac{1}{4}$	$\frac{1}{8}$	$\frac{1}{8}$	

g

$\frac{1}{8}$	$\frac{1}{2}$	$\frac{1}{4}$	

h

$\frac{1}{4}$	$\frac{1}{2}$	$\frac{1}{8}$	

2 Will each collection of fractions make exactly one strip, more than one strip or less than one strip?

a

b

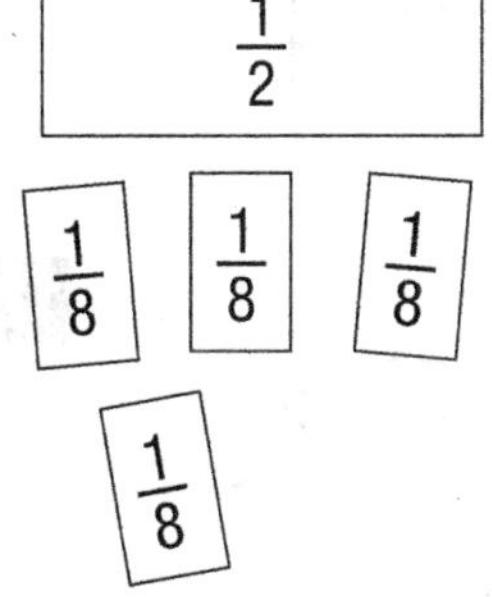

c

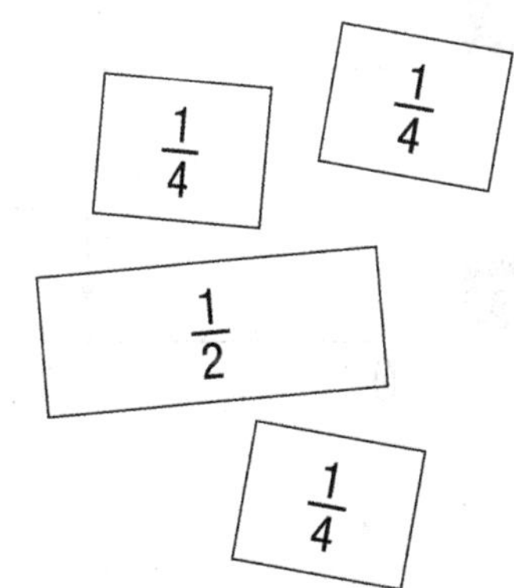

Name fractions on number lines

Remember

A mixed number is a number that has a whole number and a common fraction.

Examples include $1\frac{1}{2}$, $3\frac{4}{5}$ and $2\frac{3}{4}$.

What fraction or mixed number is shown on each number line?

1
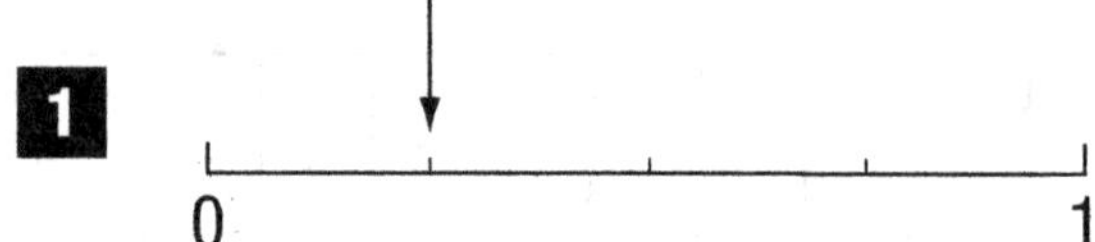

2
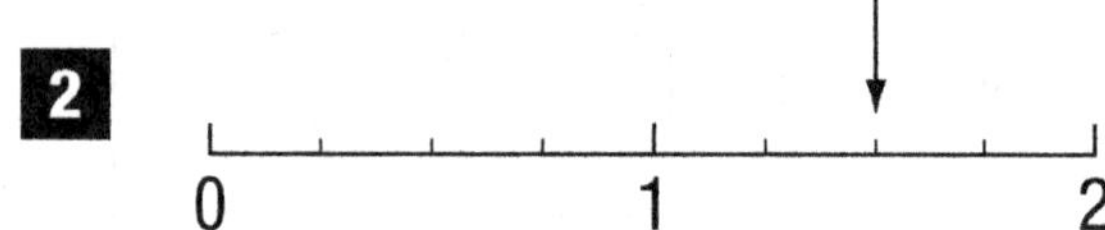

3
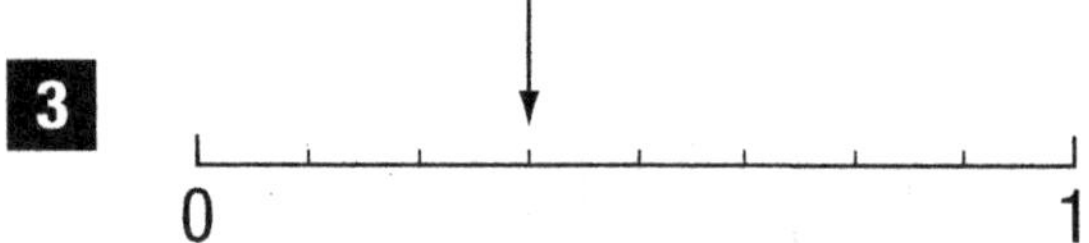

4
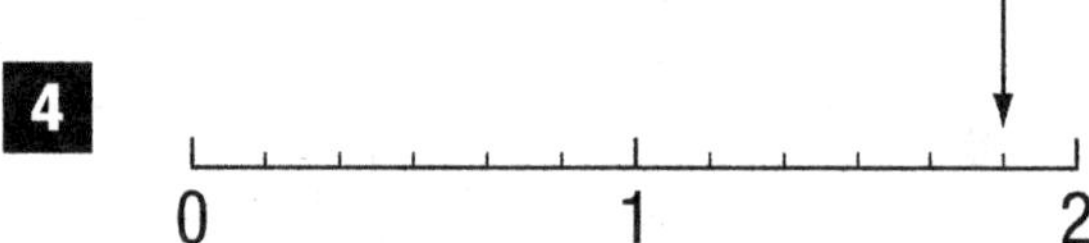

5
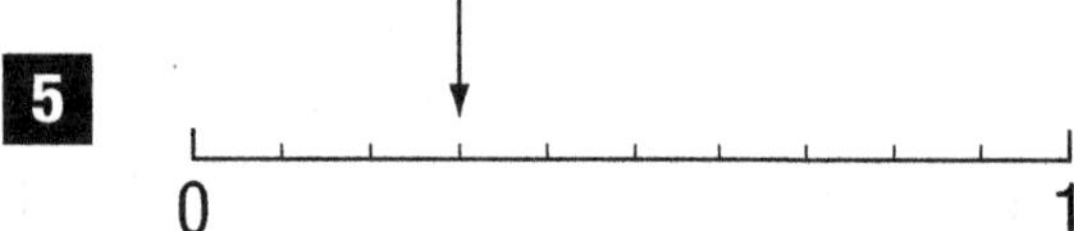

6
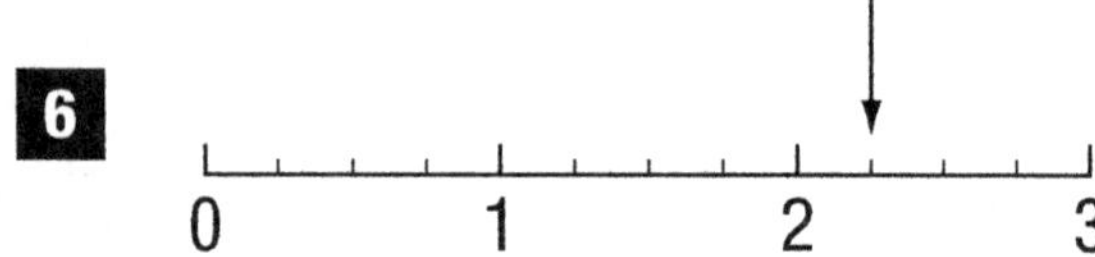

7
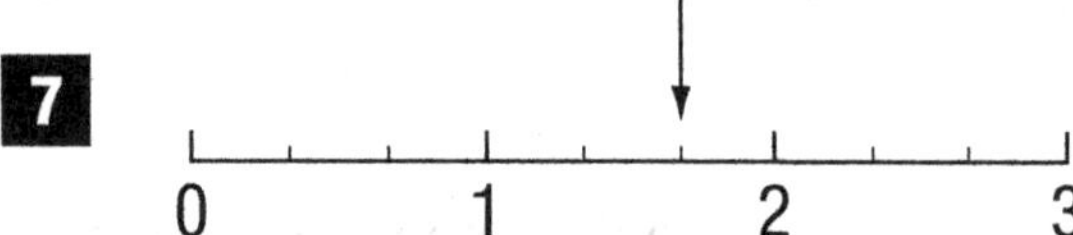

8
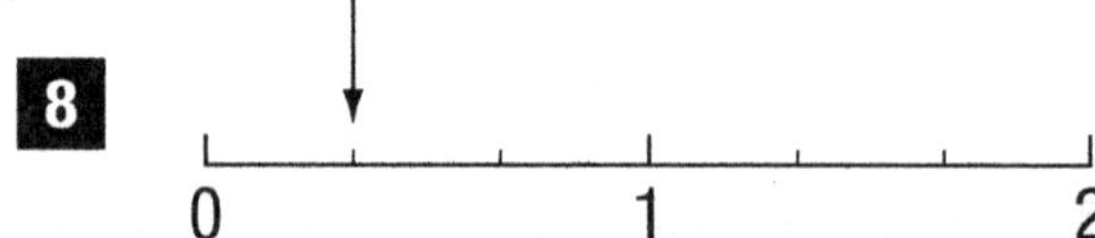

9
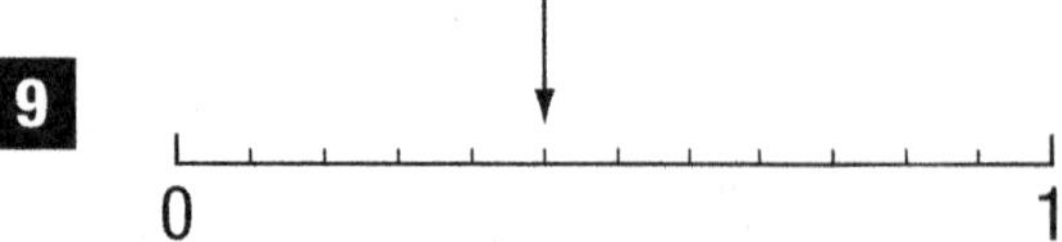

10
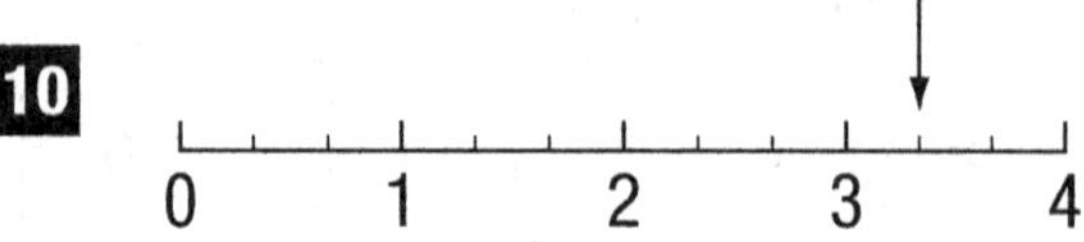

11

12
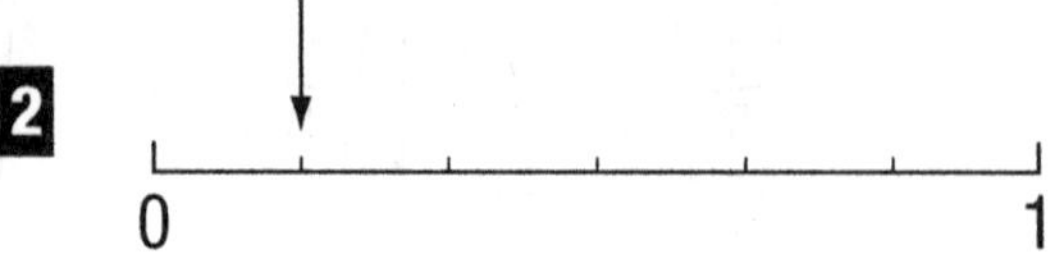

Place fractions on number lines

Draw each number line and place the fraction or mixed number shown at the start in its correct place.

1 $\frac{1}{5}$ — 0 to 1

2 $\frac{1}{3}$ — 0, 1, 2

3 $\frac{3}{4}$ — 0, 1, 2

4 $2\frac{1}{2}$ — 0, 1, 2, 3

5 $5\frac{1}{4}$ — 0, 1, 2, 3, 4, 5, 6

6 $1\frac{3}{4}$ — 0, 1, 2

7 $\frac{7}{8}$ — 0, 1, 2

8 $\frac{7}{10}$ — 0 to 1

9 $\frac{5}{6}$ — 0, 1, 2

10 $1\frac{3}{8}$ — 0, 1, 2

11 $2\frac{3}{10}$ — 0, 1, 2, 3

12 $\frac{4}{7}$ — 0, 1, 2

13 $4\frac{2}{5}$ — 0, 1, 2, 3, 4, 5

14 $2\frac{2}{3}$ — 0, 1, 2, 3

Compare fractions

Use the fraction wall to help you answer the questions on the next page.

1

$\frac{1}{2}$ | $\frac{1}{2}$

$\frac{1}{3}$ | $\frac{1}{3}$ | $\frac{1}{3}$

$\frac{1}{4}$ | $\frac{1}{4}$ | $\frac{1}{4}$ | $\frac{1}{4}$

$\frac{1}{5}$ | $\frac{1}{5}$ | $\frac{1}{5}$ | $\frac{1}{5}$ | $\frac{1}{5}$

$\frac{1}{6}$ | $\frac{1}{6}$ | $\frac{1}{6}$ | $\frac{1}{6}$ | $\frac{1}{6}$ | $\frac{1}{6}$

$\frac{1}{7}$ | $\frac{1}{7}$ | $\frac{1}{7}$ | $\frac{1}{7}$ | $\frac{1}{7}$ | $\frac{1}{7}$ | $\frac{1}{7}$

$\frac{1}{8}$ | $\frac{1}{8}$ | $\frac{1}{8}$ | $\frac{1}{8}$ | $\frac{1}{8}$ | $\frac{1}{8}$ | $\frac{1}{8}$ | $\frac{1}{8}$

$\frac{1}{9}$ | $\frac{1}{9}$ | $\frac{1}{9}$ | $\frac{1}{9}$ | $\frac{1}{9}$ | $\frac{1}{9}$ | $\frac{1}{9}$ | $\frac{1}{9}$ | $\frac{1}{9}$

$\frac{1}{10}$ | $\frac{1}{10}$ | $\frac{1}{10}$ | $\frac{1}{10}$ | $\frac{1}{10}$ | $\frac{1}{10}$ | $\frac{1}{10}$ | $\frac{1}{10}$ | $\frac{1}{10}$ | $\frac{1}{10}$

1 Write the fraction that is bigger in each pair.

a $\frac{1}{4}$ or $\frac{1}{8}$ b $\frac{1}{3}$ or $\frac{1}{5}$ c $\frac{1}{9}$ or $\frac{1}{7}$

d $\frac{1}{6}$ or $\frac{1}{10}$ e $\frac{1}{2}$ or $\frac{1}{3}$ f $\frac{1}{7}$ or $\frac{1}{4}$

g $\frac{2}{3}$ or $\frac{3}{4}$ h $\frac{5}{8}$ or $\frac{5}{6}$ i $\frac{6}{7}$ or $\frac{8}{9}$

j $\frac{3}{10}$ or $\frac{2}{5}$ k $\frac{7}{8}$ or $\frac{8}{10}$ l $\frac{4}{5}$ or $\frac{2}{3}$

2 Write the smallest fraction in each group.

a $\frac{1}{7}$ $\frac{1}{3}$ $\frac{1}{10}$ b $\frac{1}{6}$ $\frac{1}{4}$ $\frac{1}{8}$ c $\frac{1}{2}$ $\frac{1}{3}$ $\frac{2}{10}$

d $\frac{2}{7}$ $\frac{1}{2}$ $\frac{2}{3}$ e $\frac{2}{9}$ $\frac{1}{7}$ $\frac{3}{4}$ f $\frac{3}{8}$ $\frac{5}{7}$ $\frac{1}{10}$

g $\frac{2}{3}$ $\frac{4}{5}$ $\frac{5}{7}$ h $\frac{3}{10}$ $\frac{2}{8}$ $\frac{2}{7}$ i $\frac{5}{6}$ $\frac{7}{8}$ $\frac{7}{9}$

3 Write the fractions that are smaller than the fractions on the left.

a $\mathbf{\frac{1}{2}}$ $\frac{2}{5}$ $\frac{7}{8}$ $\frac{1}{3}$ $\frac{5}{7}$ $\frac{3}{10}$ $\frac{3}{4}$ $\frac{1}{8}$ $\frac{4}{9}$

b $\mathbf{\frac{3}{4}}$ $\frac{1}{3}$ $\frac{1}{2}$ $\frac{9}{10}$ $\frac{5}{6}$ $\frac{7}{8}$ $\frac{1}{8}$ $\frac{6}{9}$ $\frac{6}{7}$

c $\mathbf{\frac{2}{3}}$ $\frac{3}{4}$ $\frac{7}{10}$ $\frac{4}{5}$ $\frac{1}{2}$ $\frac{2}{7}$ $\frac{7}{9}$ $\frac{5}{8}$ $\frac{5}{6}$

Compare and order fractions

Play this game with a friend. Each player needs a set of 9 counters (each set a different colour). Each pair needs 6 small cards marked with these fractions: $\frac{1}{2}, \frac{1}{2}, \frac{1}{4}, \frac{1}{4}, \frac{3}{4}, \frac{3}{4}$. Each pair also needs 6 small cards marked with these words: *more than*, *more than*, *less than*, *less than*, *equal to*, *equal to.*

- Shuffle each set of six cards and place them upside down in two piles.
- Take turns to pick up one card from the top of each pile.
- Try to find a shape on the game board that meets the conditions shown on the two cards. Place a counter on that shape. If no shape meets the conditions or if it is already covered, miss that turn. When all cards are taken shuffle the piles again.
- Continue until all shapes are covered. The winner is the player who has covered the most shapes.

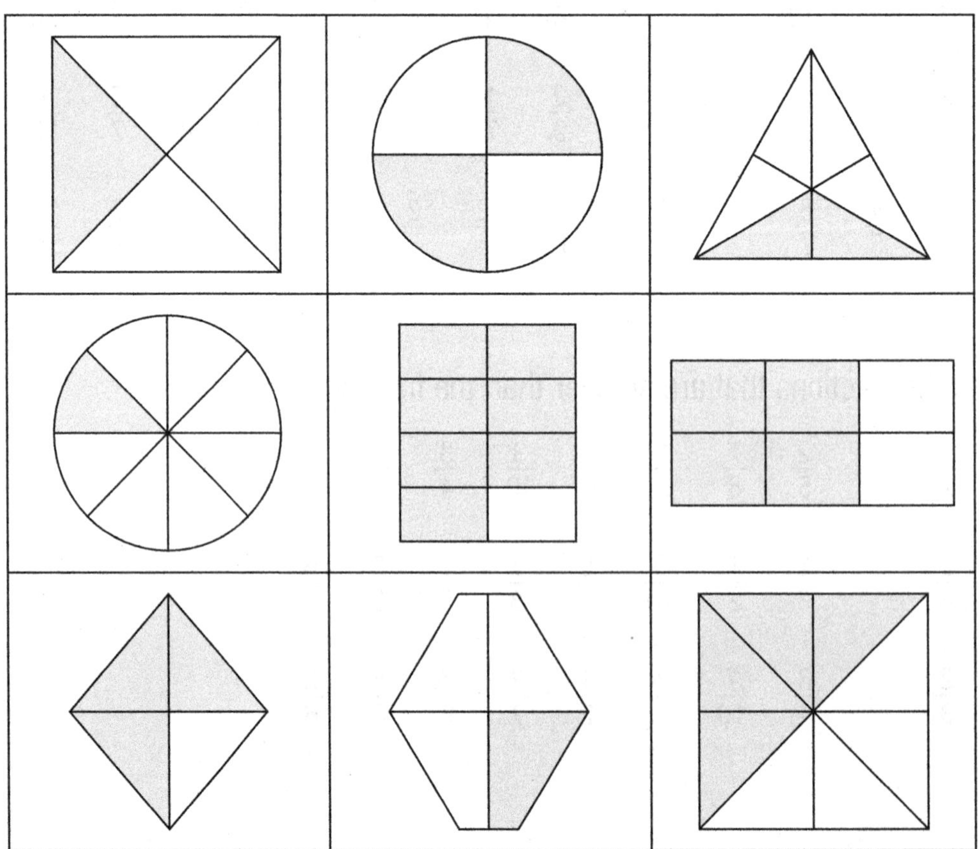

Represent common fractions

Copy and complete this chart.

	Symbol	Word	Shape	Group	Number line
1	$\frac{1}{2}$				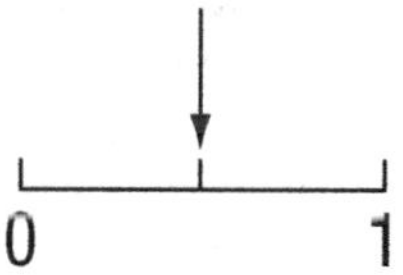 0 1
2		one quarter	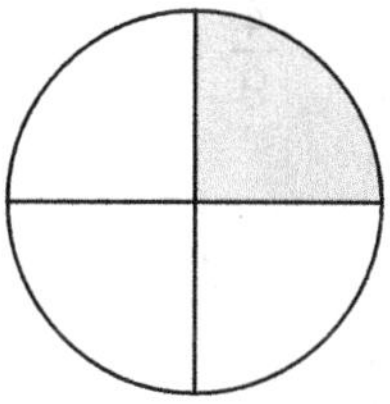		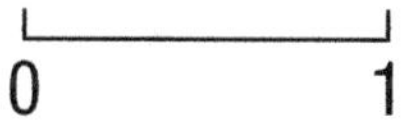 0 1
3		one eighth		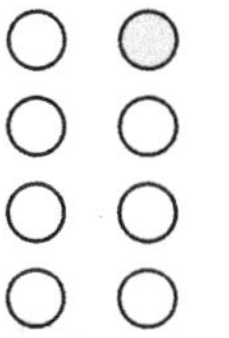	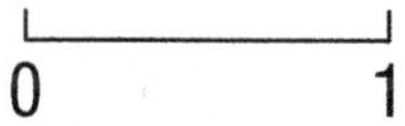 0 1
4	$\frac{1}{3}$		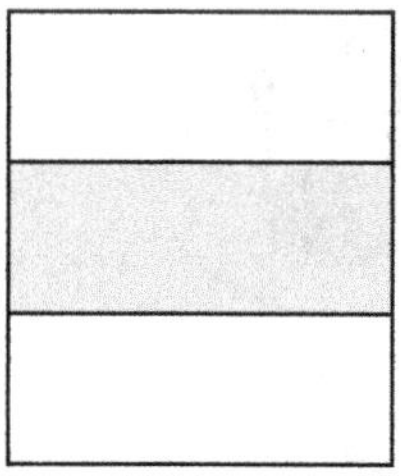		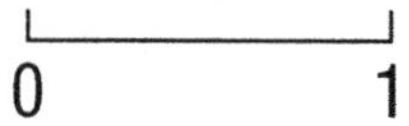 0 1
5	$\frac{1}{5}$				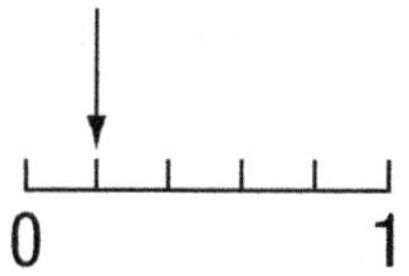 0 1
6		one tenth		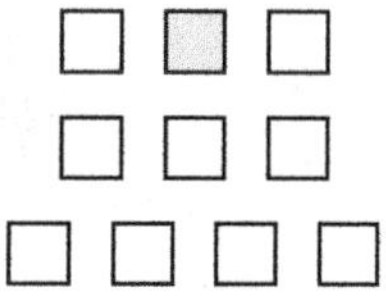	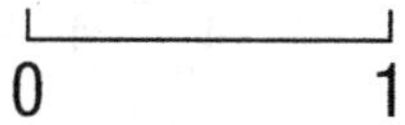 0 1

Add and subtract fractions

Set out and solve these additions and subtractions to see how many balls you can catch.

1

2

3

4

5

6

7

8

9
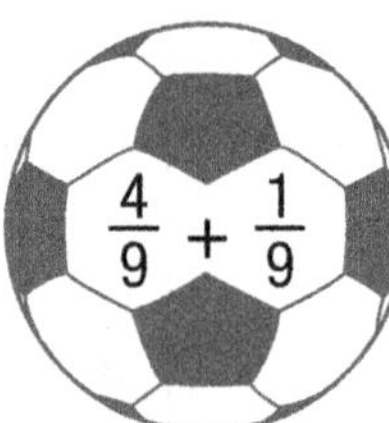

10
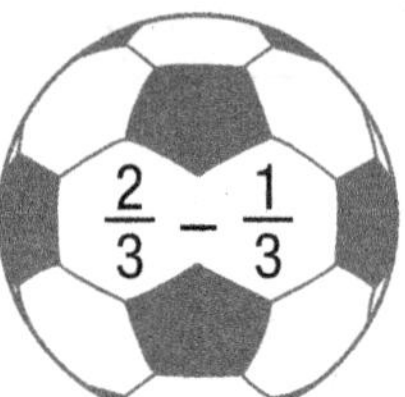

11

12

13

14

15

16

17

18

How many balls did you catch? _____

Compare and represent simple decimals up to two decimal places

Record tenths as decimals and common fractions

Write a decimal and a common fraction to show how many things in each group are shaded. The first one is done.

1

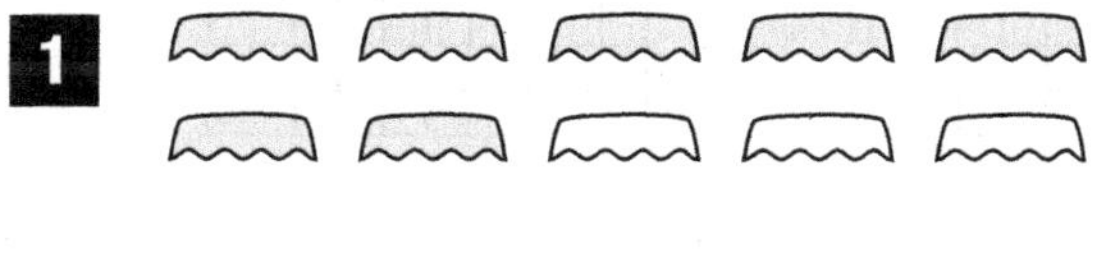

0.7 $\frac{7}{10}$

2

3

4

5

6

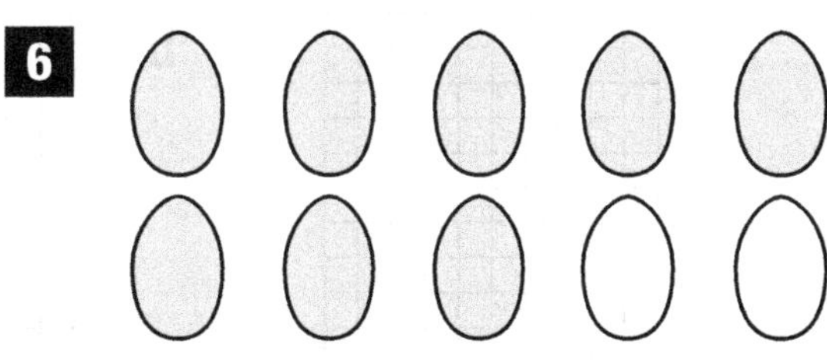

7

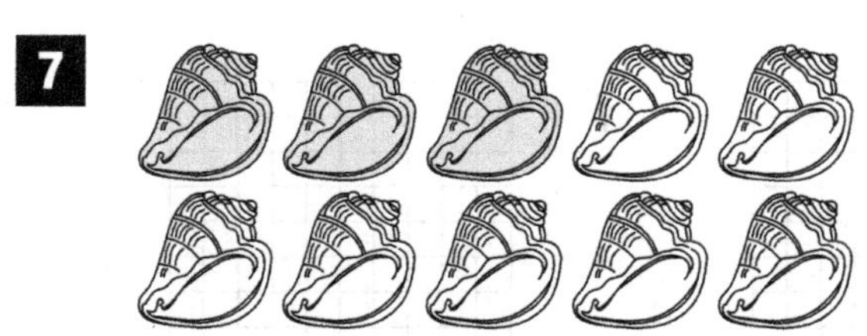

8

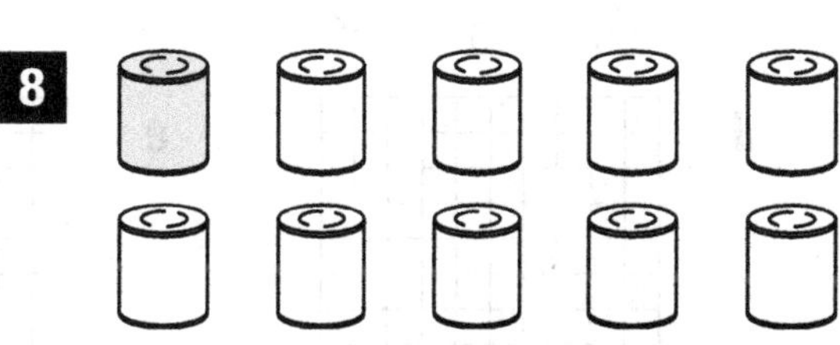

Record hundredths as decimals and common fractions

1 Write a common fraction and a decimal fraction to show how many squares are shaded in each grid. The first one is done.

a 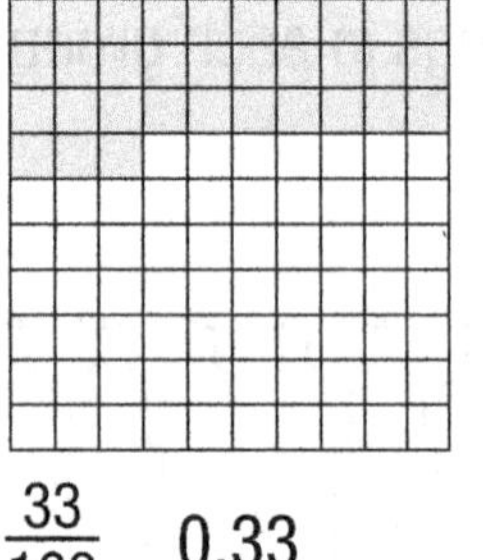

$\frac{33}{100}$ 0.33

b 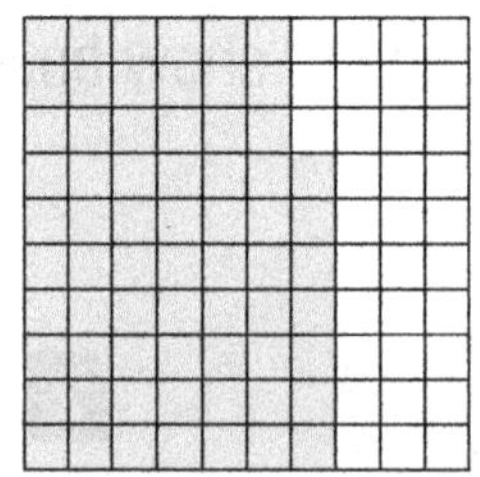

c 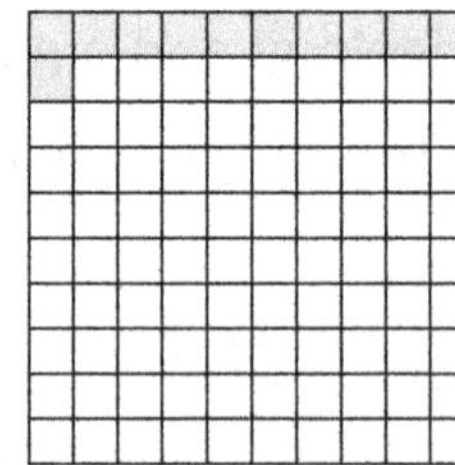

d 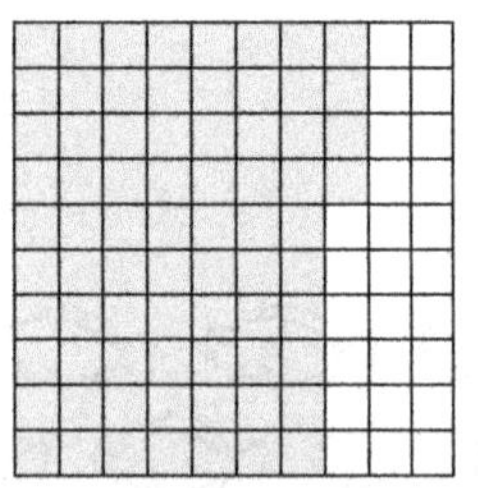

e 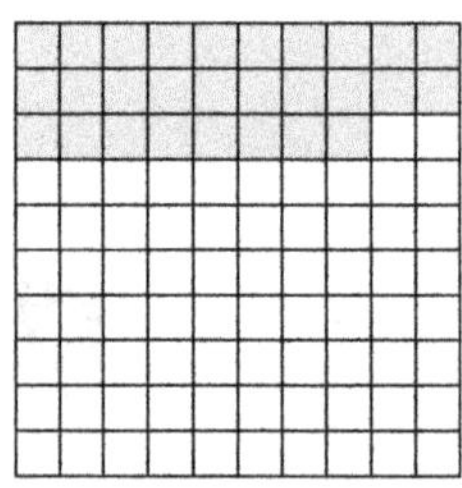

f 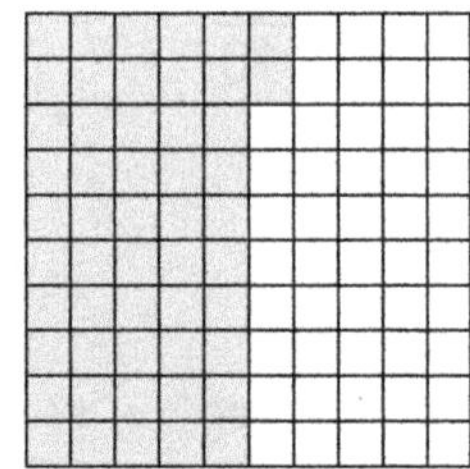

2 Write a common fraction out of 100, a common fraction out of 10 and a decimal fraction for each grid to show how many squares are shaded. The first one is done.

a 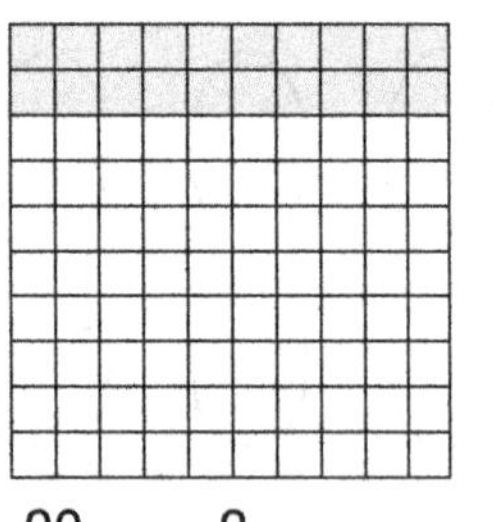

$\frac{20}{100}$ $\frac{2}{10}$ 0.2

b 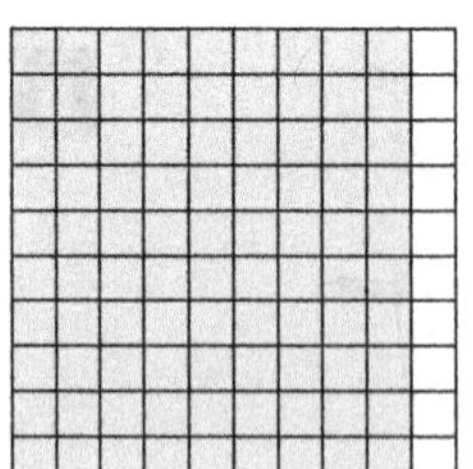

c 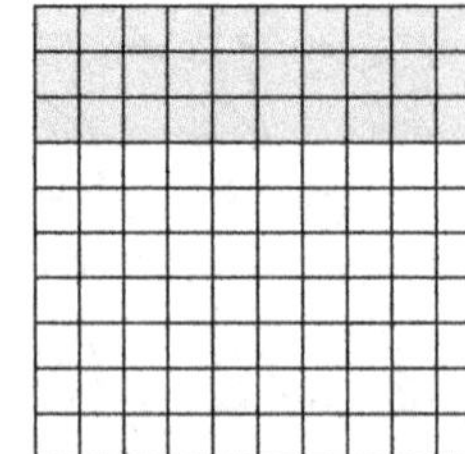

d 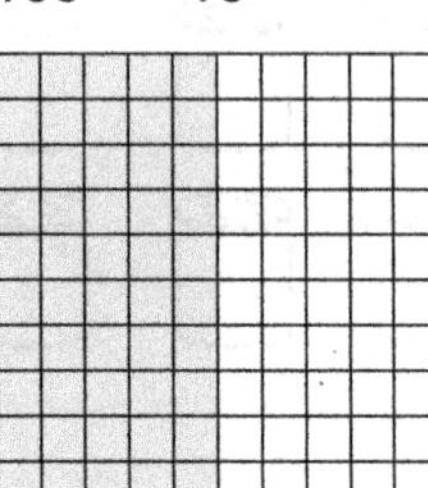

e 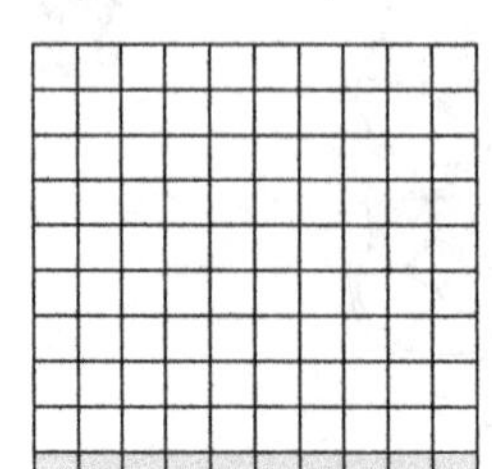

f 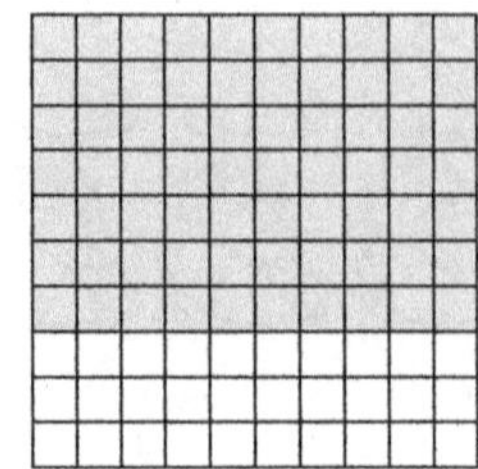

Record using decimal notation

Remember

Use a decimal point to separate the whole number part from the fraction part.

Examples: 5 tenths = 0.5; 4 and 7 tenths = 4.7; 3 and 78 hundredths = 3.78

1 Write as decimals.

a 8 tenths

b 31 hundredths

c 5 tenths

d 57 hundredths

e 2 tenths

f 28 hundredths

g 9 tenths

h 75 hundredths

i 1 tenth

j 60 hundredths

k 6 tenths

l 40 hundredths

m 4 tenths

n 12 hundredths

o 7 tenths

p 7 hundredths

q 3 tenths

r 3 hundredths

s 19 hundredths

t 8 hundredths

u 90 hundredths

v 10 hundredths

w 11 tenths

x 23 tenths

y 5 and 9 tenths

z 8 and 4 tenths

2 Write these distances as decimals.

a 2 metres 18 centimetres
b 7 metres 53 centimetres
c 1 metre 50 centimetres
d 4 metres 80 centimetres
e 5 metres 87 centimetres
f 3 metres 20 centimetres
g 145 centimetres
h 612 centimetres
i 486 centimetres
j 947 centimetres
k 84 centimetres
l 77 centimetres
m 208 centimetres
n 803 centimetres
o 506 centimetres

3 Write these amounts of money as decimals.

a 6 kina 45 toea
b 3 kina 75 toea
c 9 kina 25 toea
d 1 kina 70 toea
e 5 kina 10 toea
f 2 kina 50 toea
g 10 kina 99 toea
h 3 kina 5 toea
i 12 kina 4 toea
j 8 kina 15 toea
k 25 kina 60 toea
l 40 kina 35 toea
m 16 kina 8 toea
n 33 kina 28 toea
o 11 kina 56 toea

Interpret and record decimal notation

1 Write these common fractions as decimals.

a $\frac{6}{10}$ **b** $\frac{9}{10}$ **c** $\frac{1}{10}$ **d** $\frac{5}{10}$

e $\frac{49}{100}$ **f** $\frac{63}{100}$ **g** $\frac{21}{100}$ **h** $\frac{38}{100}$

i $\frac{30}{100}$ **j** $\frac{3}{100}$ **k** $\frac{80}{100}$ **l** $\frac{8}{100}$

m $\frac{5}{100}$ **n** $\frac{77}{100}$ **o** $\frac{4}{10}$ **p** $\frac{8}{10}$

q $\frac{29}{100}$ **r** $\frac{14}{100}$ **s** $\frac{50}{100}$ **t** $\frac{70}{100}$

u $\frac{1}{100}$ **v** $\frac{17}{100}$ **w** $\frac{87}{100}$ **x** $\frac{10}{100}$

2 Write these mixed numbers as decimals.

a $1\frac{5}{10}$ **b** $5\frac{7}{10}$ **c** $2\frac{4}{10}$ **d** $8\frac{1}{10}$

e $4\frac{36}{100}$ **f** $7\frac{84}{100}$ **g** $3\frac{49}{100}$ **h** $6\frac{70}{100}$

i $5\frac{3}{100}$ **j** $1\frac{8}{100}$ **k** $9\frac{6}{100}$ **l** $4\frac{11}{100}$

m $2\frac{7}{10}$ **n** $6\frac{9}{10}$ **o** $11\frac{3}{10}$ **p** $16\frac{2}{10}$

q $10\frac{53}{100}$ **r** $15\frac{18}{100}$ **s** $12\frac{9}{100}$ **t** $25\frac{1}{100}$

u $17\frac{21}{100}$ **v** $9\frac{40}{100}$ **w** $7\frac{50}{100}$ **x** $2\frac{5}{100}$

3 Write these decimals as common fractions or mixed numbers.

a 5.67	**b** 0.6	**c** 1.04	**d** 8.92
e 0.07	**f** 3.5	**g** 2.77	**h** 0.64
i 4.59	**j** 6.03	**k** 0.2	**l** 1.81
m 11.08	**n** 19.1	**o** 15.25	**p** 12.9

4 Write these decimal lengths as metres and centimetres.

a 2.21 metres	**b** 0.86 metres	**c** 5.7 metres
d 4.2 metres	**e** 0.9 metres	**f** 1.25 metres
g 8.5 metres	**h** 3.75 metres	**i** 10.6 metres
j 12.63 metres	**k** 9.08 metres	**l** 6.05 metres

Compare and order decimals

1 Write the larger decimal in each pair.

a 2.75 or 2.7	**b** 32.5 or 32.47	**c** 5.6 or 5.06
d 7.8 or 7.89	**e** 1.03 or 1.3	**f** 15.2 or 15.12
g 23.45 or 23.4	**h** 0.69 or 0.7	**i** 0.81 or 0.8
j 12.5 or 12.45	**k** 3.57 or 3.6	**l** 10.02 or 10.2
m 8.9 or 8.85	**n** 6.5 or 6.25	**o** 9.15 or 9.1

2 Write these decimals in order from largest to smallest

a 7.9 9.7 7.6 7.8 8.6

b 4.0 5.1 3.7 2.4 3.5

c 3.1 2.0 2.1 3.5 2.9

d 4.6 5.3 4.3 5.0 4.9

e 8.1 7.8 8.6 7.9 8.4

f 1.9 1.1 0.9 1.7 0.6

g 6.5 6.8 5.8 5.5 6.1

h 9.3 9.0 8.9 9.6 8.6

3 Write these decimals in order from smallest to largest.

a 6.79 6.7 6.85 6.67 6.72 6.07

b 1.49 1.36 1.41 1.8 1.38 1.08

c 2.12 2.01 2.2 2.19 2.09 2.1

d 3.09 3.5 3.56 2.98 3.18 3.9

e 4.95 4.09 4.9 4.99 4.89 4.19

f 0.07 0.77 0.7 0.75 0.68 0.17

g 8.56 8.06 8.59 8.6 8.07 8.63

h 5.15 5.01 5.11 5.05 5.5 5.51

i 7.23 7.32 7.3 7.02 7.35 7.29

j 9.67 9.65 9.6 9.59 9.69 9.06

k 0.04 0.4 0.39 0.41 0.14 0.47

l 2.36 2.16 2.6 2.03 2.39 2.59

Round off decimals

1 Round these lengths to the nearest whole metre (m).

a	3.8 m	**b**	1.3 m	**c**	5.7 m	**d**	2.5 m
e	1.9 m	**f**	4.2 m	**g**	6.3 m	**h**	1.8 m
i	7.5 m	**j**	2.1 m	**k**	3.4 m	**l**	5.6 m
m	9.3 m	**n**	0.7 m	**o**	1.5 m	**p**	6.4 m
q	8.1 m	**r**	4.8 m	**s**	0.3 m	**t**	0.5 m
u	10.2 m	**v**	12.6 m	**w**	11.4 m	**x**	15.9 m

2 Round these decimals to the nearest whole number.

a	5.7	**b**	2.3	**c**	7.9	**d**	6.1
e	4.25	**f**	10.83	**g**	5.78	**h**	0.37
i	1.61	**j**	8.11	**k**	3.55	**l**	9.76
m	5.09	**n**	0.95	**o**	4.21	**p**	2.65
q	11.46	**r**	8.51	**s**	1.32	**t**	6.17
u	15.04	**v**	12.67	**w**	19.25	**x**	11.59

3 Round these distances to the nearest tenth of a metre (m).

a	12.79 m	**b**	25.36 m	**c**	15.27 m	**d**	33.81 m
e	28.15 m	**f**	11.25 m	**g**	40.69 m	**h**	10.43 m
i	17.52 m	**j**	38.94 m	**k**	20.08 m	**l**	36.75 m
m	23.05 m	**n**	19.37 m	**o**	55.88 m	**p**	49.29 m
q	36.38 m	**r**	61.13 m	**s**	42.07 m	**t**	53.64 m
u	70.55 m	**v**	45.41 m	**w**	82.76 m	**x**	75.18 m

4 Round these weights to the nearest tenth of a kilogram (kg).

a	5.65 kg	**b**	3.18 kg	**c**	0.67 kg	**d**	7.34 kg
e	1.92 kg	**f**	6.59 kg	**g**	2.08 kg	**h**	11.41 kg
i	8.23 kg	**j**	0.15 kg	**k**	4.77 kg	**l**	9.84 kg
m	2.53 kg	**n**	5.92 kg	**o**	10.38 kg	**p**	6.05 kg
q	12.14 kg	**r**	8.79 kg	**s**	15.42 kg	**t**	20.86 kg
u	10.81 kg	**v**	17.54 kg	**w**	11.37 kg	**x**	19.16 kg

Add and subtract decimals (tenths)

Help Box

Add decimals:

8.5 + 6.9

```
   8.5
+ ₁6.9
  15.4
```

Subtract decimals:

9.3 − 4.7

```
  ⁸9.¹3
 − 4.7
   4.6
```

Remember

When adding and subtracting decimals, keep the decimal points under one another.

Copy and solve these decimal additions and subtractions.

1
```
  4.6
− 3.2
```

2
```
  3.5
+ 5.4
```

3
```
  6.5
− 3.4
```

4
```
  7.9
+ 2.4
```

5
```
  8.3
+ 8.8
```

6
```
  15.6
− 11.8
```

7
```
  12.7
+ 14.5
```

8
```
  9.2
− 5.7
```

9
```
  14.3
− 10.6
```

10
```
  16.8
+ 13.6
```

11
```
  8.1
− 5.4
```

12
```
  7.8
+ 6.9
```

13
```
  11.5
+ 17.8
```

14
```
  18.4
− 12.7
```

15
```
  23.6
+ 17.6
```

16
```
  19.2
− 10.5
```

Add and subtract decimals (hundredths)

Help Box

Add decimals:

$7.84 + 5.49$

$$\begin{array}{r} 7.84 \\ +\ {}^{1}5.{}^{1}49 \\ \hline 13.33 \end{array}$$

Subtract decimals:

$8.53 - 5.66$

$$\begin{array}{r} {}^{7}\not{8}.{}^{14}\not{5}{}^{1}3 \\ -\ 5.66 \\ \hline 2.87 \end{array}$$

Calculate the answers and then write the letters from the boxes that have answers less than 6. Unscramble the letters to make the name of a shape.

1 Y $\begin{array}{r} 4.42 \\ +5.11 \\ \hline \_\_\_\_ \end{array}$	**2** A $\begin{array}{r} 8.38 \\ -3.23 \\ \hline \_\_\_\_ \end{array}$	**3** B $\begin{array}{r} 3.51 \\ +6.04 \\ \hline \_\_\_\_ \end{array}$	**4** S $\begin{array}{r} 9.82 \\ -2.46 \\ \hline \_\_\_\_ \end{array}$
5 L $\begin{array}{r} 5.87 \\ -1.94 \\ \hline \_\_\_\_ \end{array}$	**6** W $\begin{array}{r} 4.62 \\ +9.71 \\ \hline \_\_\_\_ \end{array}$	**7** T $\begin{array}{r} 9.23 \\ -4.07 \\ \hline \_\_\_\_ \end{array}$	**8** M $\begin{array}{r} 8.32 \\ +0.91 \\ \hline \_\_\_\_ \end{array}$
9 P $\begin{array}{r} 6.67 \\ +8.86 \\ \hline \_\_\_\_ \end{array}$	**10** G $\begin{array}{r} 8.29 \\ -2.37 \\ \hline \_\_\_\_ \end{array}$	**11** C $\begin{array}{r} 7.15 \\ +2.98 \\ \hline \_\_\_\_ \end{array}$	**12** N $\begin{array}{r} 6.05 \\ -1.89 \\ \hline \_\_\_\_ \end{array}$
13 E $\begin{array}{r} 4.11 \\ -2.92 \\ \hline \_\_\_\_ \end{array}$	**14** H $\begin{array}{r} 8.14 \\ +2.39 \\ \hline \_\_\_\_ \end{array}$	**15** R $\begin{array}{r} 9.21 \\ -5.48 \\ \hline \_\_\_\_ \end{array}$	**16** I $\begin{array}{r} 1.29 \\ +4.57 \\ \hline \_\_\_\_ \end{array}$

What is the name of the shape? ____________________

Multiply decimals

Remember

When multiplying decimals, calculate as a normal multiplication and then estimate by rounding to help you place the decimal point in the answer.

Where would you place the decimal point in each answer?

1 $\begin{array}{r} 3.2 \\ \times 3 \\ \hline 96 \end{array}$

2 $\begin{array}{r} 2.2 \\ \times 4 \\ \hline 88 \end{array}$

3 $\begin{array}{r} 5.3 \\ \times 3 \\ \hline 159 \end{array}$

4 $\begin{array}{r} 1.5 \\ \times 5 \\ \hline 75 \end{array}$

5 $\begin{array}{r} 4.7 \\ \times 6 \\ \hline 282 \end{array}$

6 $\begin{array}{r} 3.6 \\ \times 8 \\ \hline 288 \end{array}$

7 $\begin{array}{r} 2.8 \\ \times 7 \\ \hline 196 \end{array}$

8 $\begin{array}{r} 6.4 \\ \times 4 \\ \hline 256 \end{array}$

9 $\begin{array}{r} 0.8 \\ \times 9 \\ \hline 72 \end{array}$

10 $\begin{array}{r} 7.5 \\ \times 7 \\ \hline 525 \end{array}$

11 $\begin{array}{r} 11.6 \\ \times 5 \\ \hline 580 \end{array}$

12 $\begin{array}{r} 13.4 \\ \times 3 \\ \hline 402 \end{array}$

13 $\begin{array}{r} 5.63 \\ \times 4 \\ \hline 2252 \end{array}$

14 $\begin{array}{r} 3.17 \\ \times 6 \\ \hline 1902 \end{array}$

15 $\begin{array}{r} 7.32 \\ \times 7 \\ \hline 5124 \end{array}$

16 $\begin{array}{r} 4.88 \\ \times 4 \\ \hline 1952 \end{array}$

17 $\begin{array}{r} 2.76 \\ \times 8 \\ \hline 2208 \end{array}$

18 $\begin{array}{r} 6.08 \\ \times 9 \\ \hline 5472 \end{array}$

19 $\begin{array}{r} 8.29 \\ \times 5 \\ \hline 4145 \end{array}$

20 $\begin{array}{r} 3.64 \\ \times 7 \\ \hline 2548 \end{array}$

Divide decimals

Remember

When dividing decimals, calculate as a normal division and then estimate by rounding to help you place the decimal point in the answer.

1 Where would you place the decimal point in each answer?

a $\begin{array}{r} 2\,2 \\ 4\overline{)8.8} \end{array}$ b $\begin{array}{r} 1\,3 \\ 5\overline{)6.5} \end{array}$ c $\begin{array}{r} 2\,4 \\ 4\overline{)9.6} \end{array}$ d $\begin{array}{r} 1\,7 \\ 3\overline{)5.1} \end{array}$

e $\begin{array}{r} 3\,4 \\ 6\overline{)20.4} \end{array}$ f $\begin{array}{r} 2\,7 \\ 8\overline{)21.6} \end{array}$ g $\begin{array}{r} 9\,5 \\ 3\overline{)28.5} \end{array}$ h $\begin{array}{r} 3\,8 \\ 4\overline{)15.2} \end{array}$

i $\begin{array}{r} 2\,5\,3 \\ 5\overline{)12.65} \end{array}$ j $\begin{array}{r} 3\,1\,6 \\ 7\overline{)22.12} \end{array}$ k $\begin{array}{r} 1\,8\,4 \\ 6\overline{)11.04} \end{array}$ l $\begin{array}{r} 4\,7\,7 \\ 8\overline{)38.16} \end{array}$

2 Match each division with its answer.

a $4\overline{)24.8}$ b $6\overline{)28.8}$

c $2\overline{)36.8}$ d $3\overline{)26.7}$

e $5\overline{)19.25}$ f $7\overline{)35.91}$

g $3\overline{)25.71}$ h $8\overline{)18.48}$

i $6\overline{)11.22}$ j $4\overline{)29.12}$

2.31 18.4 5.13

7.28 3.85

8.57 6.2 8.9

1.87 4.8

Solve decimal word problems

Decide what process to use and then solve each word problem.

1 Philip walks 2.3 kilometres to get to school. Peter walks 0.8 kilometres. How much further does Philip walk than Peter?

2 If one bucket holds 5.6 litres of water and another bucket holds 7.8 litres of water, how much water do they hold in total?

3 Elsie uses 2.7 metres of fabric to make a dress. How much fabric will she need to make the following number of dresses?

a 5 dresses

b 3 dresses

c 8 dresses

4 The width of a school garden is 17.5 metres. The length of the same garden is 24.2 metres. What is the difference between the width and length measurements of the garden?

5 Mrs Kariko spent K8.45 at the trade store on Monday, K5.80 on Tuesday and K7.65 on Wednesday.

a How much did she spend over the three days?

b How much more did she spend on Wednesday than on Tuesday?

6 A tourist has two bags, one weighing 15.7 kilograms and the other weighing 12.8 kilograms. What is the total weight of the two bags?

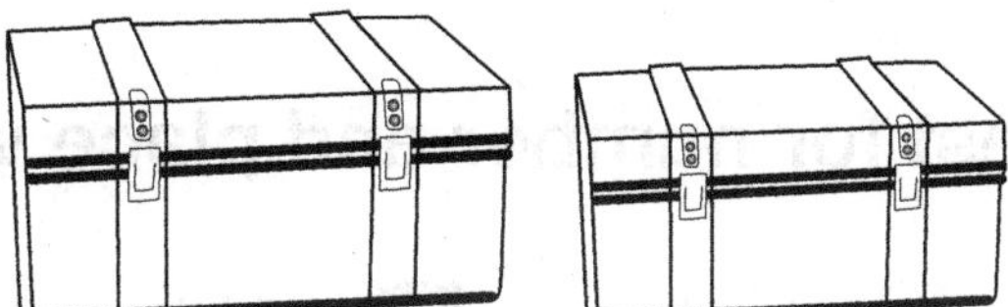

7 Naomi walks 7.8 kilometres over three days. If she walks the same distance each day, how far is this?

8 An item at the trade store costs K3.85. If Josephine buys four of the same item, how much will she spend?

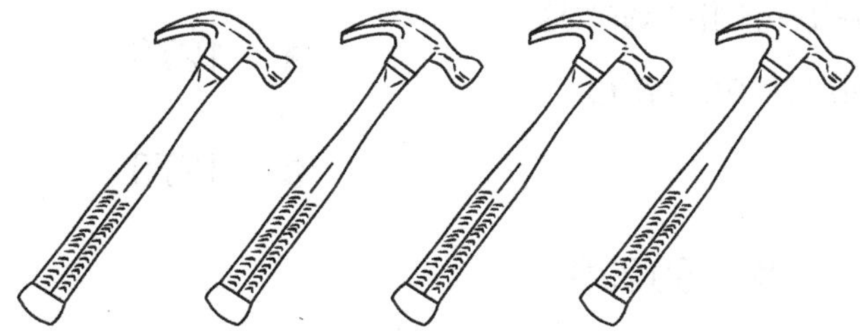

Assessment **Number and Application**

Multiple choice test for number and place value (4.1.1)

1 Which number matches the words *six hundred and fifty-eight?*

a 6058 **b** 658

c 568 **d** 60 058

2 Which number matches the words *five thousand, two hundred and thirty-nine?*

a 52 039 **b** 50 239

c 5239 **d** 5209

3 Which number is more than 2098?

a 2030 **b** 2068

c 2100 **d** 2090

4 Which number is less than 8115?

a 8120 **b** 8118

c 8200 **d** 8110

5 Which number is largest?

a 764 **b** 746

c 740 **d** 760

6 Which number is smallest?

a 1276 **b** 1260

c 1270 **d** 1267

7 Which number is between 380 and 415?

a 418 **b** 430

c 370 **d** 407

8 Which number is between 7040 and 7090?

a 7400 **b** 7055

c 7035 **d** 7006

9 What number is 2 tens, 8 ones, 6 thousands and 1 hundred?

a 6128 b 1628

c 6218 d 6182

10 What is the largest number you can make with 5, 0 and 7?

a 705 b 570

c 750 d 507

11 What is the largest number you can make with 6, 8, 1 and 5?

a 8651 b 6815

c 8615 d 6851

12 What is the smallest number you can make with 4, 3 and 6?

a 364 b 436

c 643 d 346

13 What is the smallest number you can make with 5, 0, 2 and 7?

a 2750 b 2057

c 2507 d 2705

14 Which number has 7 tens?

a 710 b 170

c 107 d 701

15 Which number has 2 hundreds?

a 1325 b 2153

c 5213 d 3125

16 What is the next number? 573, 580, 587, 594, ____

a 601 b 599

c 600 d 602

Assessment Number and Application

Operations (4.1.2)

1 Add each set of numbers in your head and write the answers.

a 6 + 9 + 3 + 4 + 7 + 5 **b** 8 + 8 + 5 + 7 + 4 + 6

c 4 + 6 + 6 + 3 + 9 + 8 **d** 7 + 7 + 8 + 9 + 8 + 9

2 Add each pair of numbers in your head and write the answers.

a 30 + 50 **b** 70 + 20 **c** 40 + 40 **d** 60 + 50

e 20 + 48 **f** 39 + 50 **g** 29 + 29 **h** 45 + 39

3 Subtract each pair of numbers in your head and write the answers.

a 15 – 6 **b** 17 – 5 **c** 13 – 8 **d** 16 – 9

e 18 – 9 **f** 20 – 11 **g** 19 – 7 **h** 12 – 5

i 32 – 8 **j** 26 – 9 **k** 35 – 7 **l** 23 – 6

m 41 – 9 **n** 43 – 8 **o** 51 – 6 **p** 44 – 7

4 Multiply these numbers in your head and write the answers.

a	3×4	**b**	4×9	**c**	11×7	**d**	6×8
e	6×12	**f**	7×3	**g**	5×6	**h**	12×4
i	3×9	**j**	6×7	**k**	4×7	**l**	11×11

5 Divide these numbers in your head and write the answers.

a	$27 \div 3$	**b**	$54 \div 6$	**c**	$24 \div 4$	**d**	$66 \div 11$
e	$36 \div 6$	**f**	$110 \div 11$	**g**	$18 \div 3$	**h**	$32 \div 4$
i	$77 \div 11$	**j**	$16 \div 4$	**k**	$24 \div 6$	**l**	$21 \div 3$

6 Multiply these numbers by 10 and write the answers.

a	13×10	**b**	47×10	**c**	24×10	**d**	61×10
e	352×10	**f**	125×10	**g**	268×10	**h**	543×10

7 Write the numbers that are multiples of each number on the left.

a **6** 18 7 23 30 12 40 24 60 32 19

b **9** 25 45 36 17 24 48 27 39 63 81

c **4** 12 22 40 48 9 16 36 25 28 34

8 Solve these additions.

a $\begin{array}{r} 433 \\ +\underline{564} \\ \underline{} \end{array}$ b $\begin{array}{r} 286 \\ +\underline{375} \\ \underline{} \end{array}$ c $\begin{array}{r} 3106 \\ +\underline{5673} \\ \underline{} \end{array}$ d $\begin{array}{r} 2566 \\ +\underline{5848} \\ \underline{} \end{array}$

9 Solve these subtractions.

a $\begin{array}{r} 766 \\ -\underline{214} \\ \underline{} \end{array}$ b $\begin{array}{r} 627 \\ -\underline{458} \\ \underline{} \end{array}$ c $\begin{array}{r} 9784 \\ -\underline{4052} \\ \underline{} \end{array}$ d $\begin{array}{r} 6134 \\ -\underline{3857} \\ \underline{} \end{array}$

10 Solve these multiplications.

a $\begin{array}{r} 232 \\ \underline{\times 2} \\ \underline{} \end{array}$ b $\begin{array}{r} 458 \\ \underline{\times 7} \\ \underline{} \end{array}$ c $\begin{array}{r} 3125 \\ \underline{\times 2} \\ \underline{} \end{array}$ d $\begin{array}{r} 1768 \\ \underline{\times 5} \\ \underline{} \end{array}$

11 Solve these divisions.

a $5\overline{)745}$ b $4\overline{)632}$ c $3\overline{)4701}$ d $8\overline{)3144}$

12 Solve these divisions that have remainders.

a $5\overline{)617}$ b $3\overline{)455}$ c $7\overline{)4451}$ d $4\overline{)5817}$

13 Solve these word problems.

a Paul had K2715 in his bank account. He spent K1288 on a holiday. How much was left in his bank account?

b A school has two water tanks. One has 2565 litres and the other has 1876 litres. What is the total amount of water in the two tanks?

c A truck transports products weighing 535 kilograms every trip. What weight of products is transported over 7 trips?

Assessment — Number and Application

Fractions (4.1.3)

1 What fraction of each shape or collection has been shaded?

a 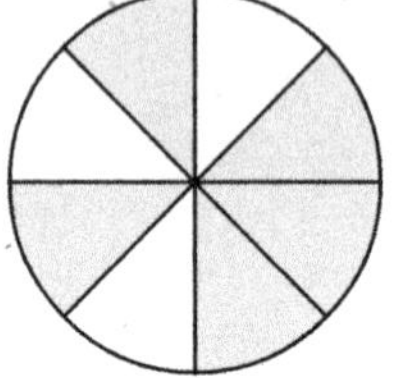

b 

c 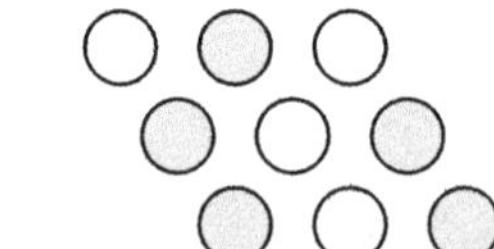

d 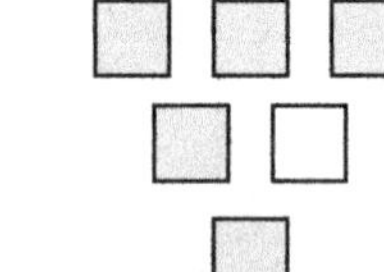

2 Which of these fraction collections equals 1?

a 

b 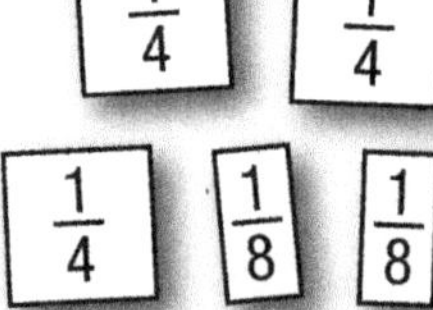

c $\frac{1}{4}$ $\frac{1}{8}$ $\frac{1}{4}$ $\frac{1}{2}$

3 Write the fraction or mixed number that is shown on each number line.

a

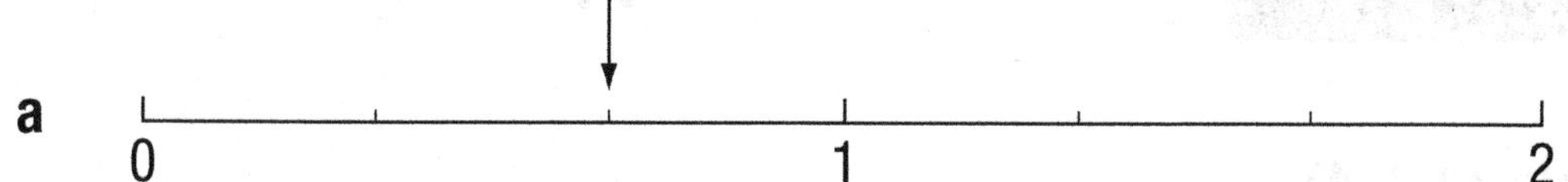

b

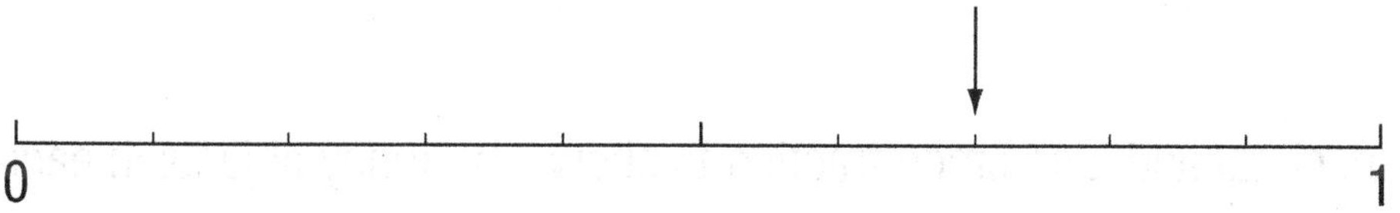

0 1

c

0 1 2 3

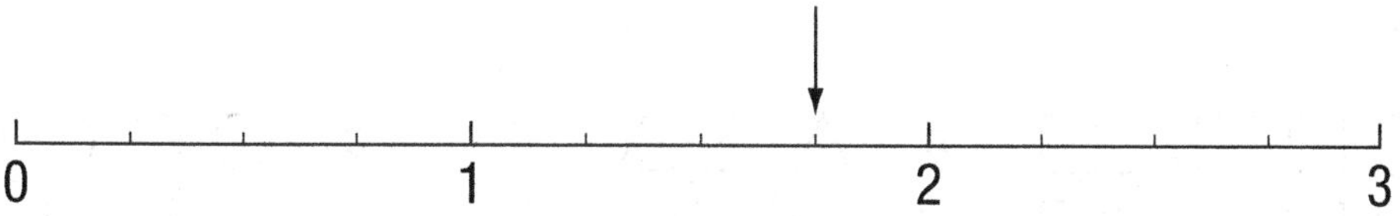

4 Which of these fractions are smaller than $\frac{1}{2}$?

$\frac{3}{4}$ $\frac{1}{4}$ $\frac{1}{3}$ $\frac{5}{6}$ $\frac{9}{10}$ $\frac{2}{7}$ $\frac{1}{8}$ $\frac{8}{9}$

5 Solve these fraction additions.

a $\frac{5}{9} + \frac{2}{9}$ **b** $\frac{5}{10} + \frac{4}{10}$ **c** $\frac{1}{6} + \frac{4}{6}$ **d** $\frac{2}{7} + \frac{3}{7}$

6 Solve these fraction subtractions.

a $\frac{9}{10} - \frac{2}{10}$ **b** $\frac{11}{12} - \frac{6}{12}$ **c** $\frac{8}{9} - \frac{4}{9}$ **d** $\frac{5}{8} - \frac{2}{8}$

Decimals (4.1.4)

1 Write a decimal and a common fraction to show how many objects in each group are shaded.

a 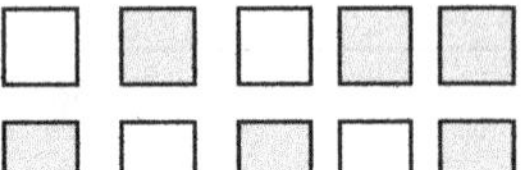 **b**  **c** 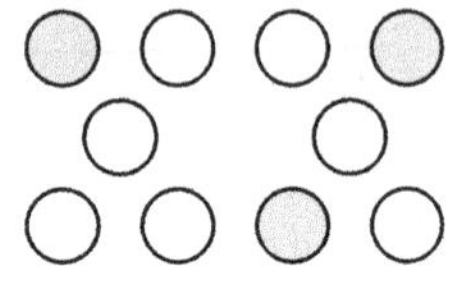

2 Write as decimals.

a 7 tenths **b** 43 hundredths **c** 3 tenths

d 59 hundredths **e** 5 hundredths **f** 80 hundredths

3 Write these lengths as decimals.

a 3 metres 27 centimetres **b** 2 metres 41 centimetres

c 5 metres 6 centimetres **d** 1 metre 50 centimetres

e 917 centimetres **f** 78 centimetres

4 Write these common fractions and mixed numbers as decimals.

a $\frac{4}{10}$ b $\frac{8}{10}$ c $\frac{61}{100}$ d $\frac{44}{100}$

e $\frac{30}{100}$ f $8\frac{7}{10}$ g $6\frac{33}{100}$ h $3\frac{5}{100}$

5 Write these decimals as common fractions or mixed numbers.

a 0.8 b 2.13 c 1.09 d 0.03

e 4.63 f 11.2 g 3.87 h 5.9

6 Write the larger decimal in each pair.

a 0.08 or 0.8 b 5.6 or 5.16 c 2.39 or 2.4 d 6.7 or 6.79

e 1.5 or 1.05 f 8.07 or 8.1 g 3.6 or 3.59 h 7.18 or 7.2

7 Round these decimals to the nearest whole number.

a 4.6 b 2.3 c 0.7 d 8.5

e 1.59 f 3.08 g 9.61 h 6.22

8 Round these decimals to the nearest tenth.

a 5.83 b 2.09 c 4.55 d 8.17

e 0.42 f 1.75 g 7.86 h 3.05

9 Solve these decimal additions.

a $\begin{array}{r} 4.6 \\ +7.9 \\ \hline \end{array}$ ____

b $\begin{array}{r} 11.5 \\ +13.8 \\ \hline \end{array}$ ____

c $\begin{array}{r} 5.34 \\ +6.87 \\ \hline \end{array}$ ____

d $\begin{array}{r} 8.09 \\ +4.75 \\ \hline \end{array}$ ____

10 Solve these decimal subtractions.

a $\begin{array}{r} 9.5 \\ -4.7 \\ \hline \end{array}$ ____

b $\begin{array}{r} 15.2 \\ -11.5 \\ \hline \end{array}$ ____

c $\begin{array}{r} 8.17 \\ -3.58 \\ \hline \end{array}$ ____

d $\begin{array}{r} 9.33 \\ -5.79 \\ \hline \end{array}$ ____

11 Where would you place the decimal point in each answer?

a $\begin{array}{r} 2.8 \\ \times 6 \\ \hline 1\,6\,8 \end{array}$

b $\begin{array}{r} 12.4 \\ \times 8 \\ \hline 9\,9\,2 \end{array}$

c $\begin{array}{r} 5.37 \\ \times 5 \\ \hline 2\,6\,8\,5 \end{array}$

d $\begin{array}{r} 9.65 \\ \times 7 \\ \hline 6\,7\,5\,5 \end{array}$

e $3\overline{)14.1}$ answer: 4 7

f $5\overline{)31.5}$ answer: 6 3

g $7\overline{)95.2}$ answer: 1 3 6

h $4\overline{)90.8}$ answer: 2 2 7

12 Solve these word problems.

a One bag of groceries weighs 2.7 kilograms and another bag weighs 4.5 kilograms. What is the total weight of the two bags?

b The same game cost K7.35 at one store and K9.10 at another store. What is the difference in price between the two stores?

Strand **Measurement**

Estimate and measure lengths, distances and perimeters using standard units of length

Estimate, measure and record length

Draw a chart like the one below. Estimate the length of each line in centimetres (cm) and record it in the chart. Then measure each line and record its length in the chart as shown. The first one has been done for you.

	Estimate	Length in cm and mm	Length in mm
1.	10 cm	10 cm 5 mm	105 mm

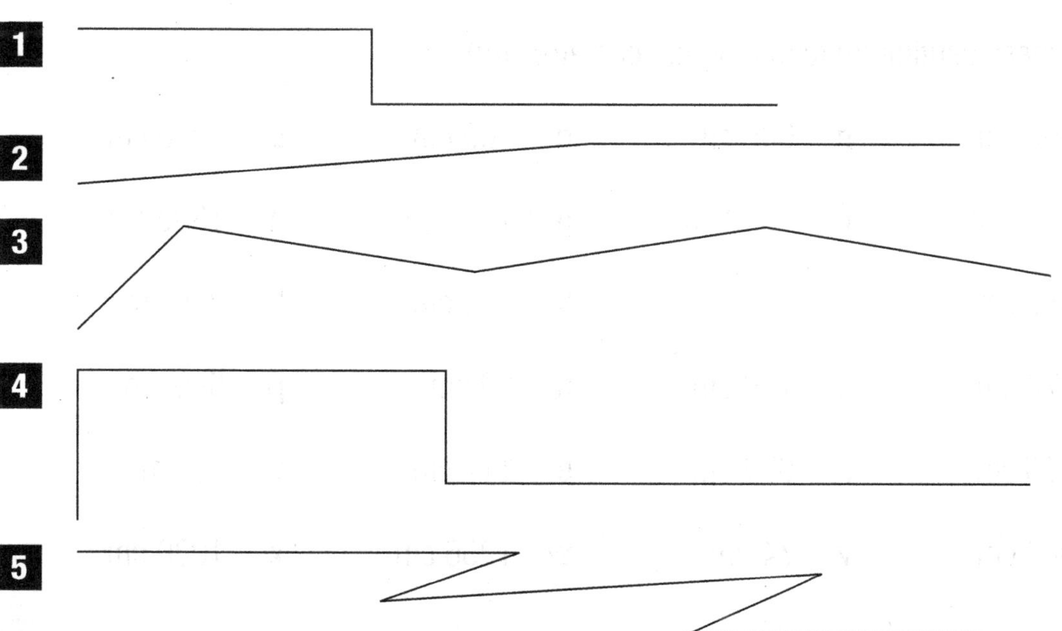

Convert length units

1 Write these metre (m) lengths as centimetres (cm).

a 3 m	**b** 5 m	**c** 2 m	**d** 8 m
e 1.5 m	**f** 4.5 m	**g** $6\frac{1}{2}$ m	**h** $9\frac{1}{2}$ m
i 5.4 m	**j** 3.6 m	**k** 1.7 m	**l** 7.3 m
m 2.25 m	**n** 1.65 m	**o** 4.12 m	**p** 0.78 m
q 3.41 m	**r** 5.72 m	**s** 0.56 m	**t** 1.87 m
u 8.07 m	**v** 10.39 m	**w** 12.6 m	**x** 11.4 m

2 Write these centimetre (cm) lengths as metres (m).

a 200 cm	**b** 600 cm	**c** 100 cm	**d** 1000 cm
e 1500 cm	**f** 1800 cm	**g** 1100 cm	**h** 1300 cm
i 350 cm	**j** 750 cm	**k** 450 cm	**l** 850 cm
m 270 cm	**n** 120 cm	**o** 90 cm	**p** 540 cm
q 325 cm	**r** 672 cm	**s** 117 cm	**t** 39 cm
u 843 cm	**v** 74 cm	**w** 1256 cm	**x** 1080 cm

3 Write these centimetre (cm) lengths as millimetres (mm).

a	8 cm	**b**	3 cm	**c**	5 cm	**d**	1 cm
e	4.5 cm	**f**	2.5 cm	**g**	$7\frac{1}{2}$ cm	**h**	$10\frac{1}{2}$ cm
i	0.6 cm	**j**	1.4 cm	**k**	6.8 cm	**l**	5.1 cm
m	2.6 cm	**n**	8.3 cm	**o**	4.9 cm	**p**	3.2 cm
q	1.7 cm	**r**	9.9 cm	**s**	7.4 cm	**t**	0.3 cm
u	15.2 cm	**v**	20.5 cm	**w**	12.8 cm	**x**	10.9 cm

4 Write these millimetre (mm) lengths as centimetres (cm).

a	60 mm	**b**	90 mm	**c**	150 mm	**d**	40 mm
e	120 mm	**f**	200 mm	**g**	170 mm	**h**	30 mm
i	45 mm	**j**	85 mm	**k**	105 mm	**l**	215 mm
m	38 mm	**n**	21 mm	**o**	57 mm	**p**	94 mm
q	364 mm	**r**	208 mm	**s**	635 mm	**t**	476 mm
u	1055 mm	**v**	1302 mm	**w**	1169 mm	**x**	1579 mm

Measure perimeter

Measure the perimeter of each shape in centimetres.

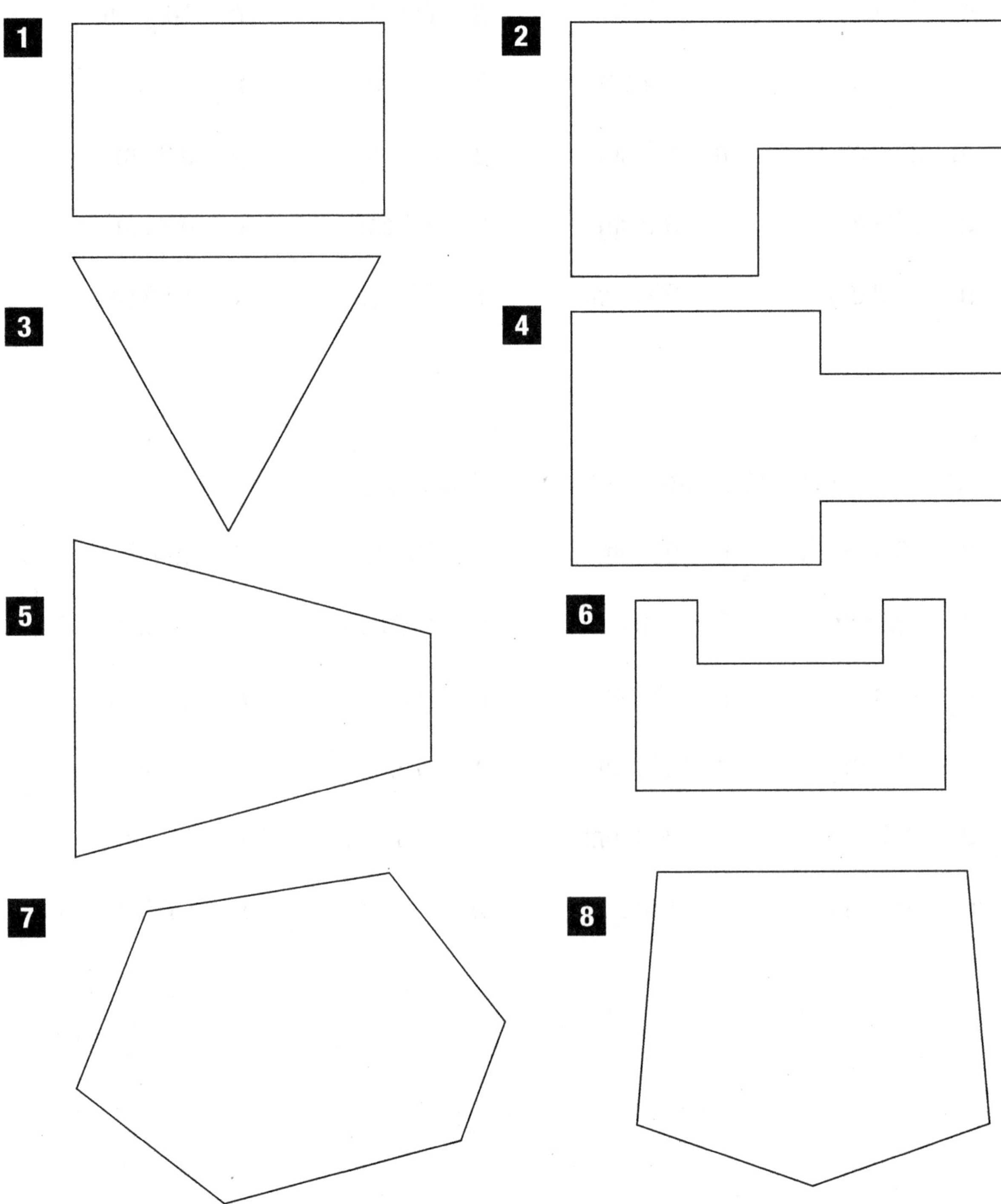

Estimate and measure areas of surfaces using standard units of area

Find area of regular and irregular shapes

These shapes are drawn on centimetre square grids. Find the area of each shape.

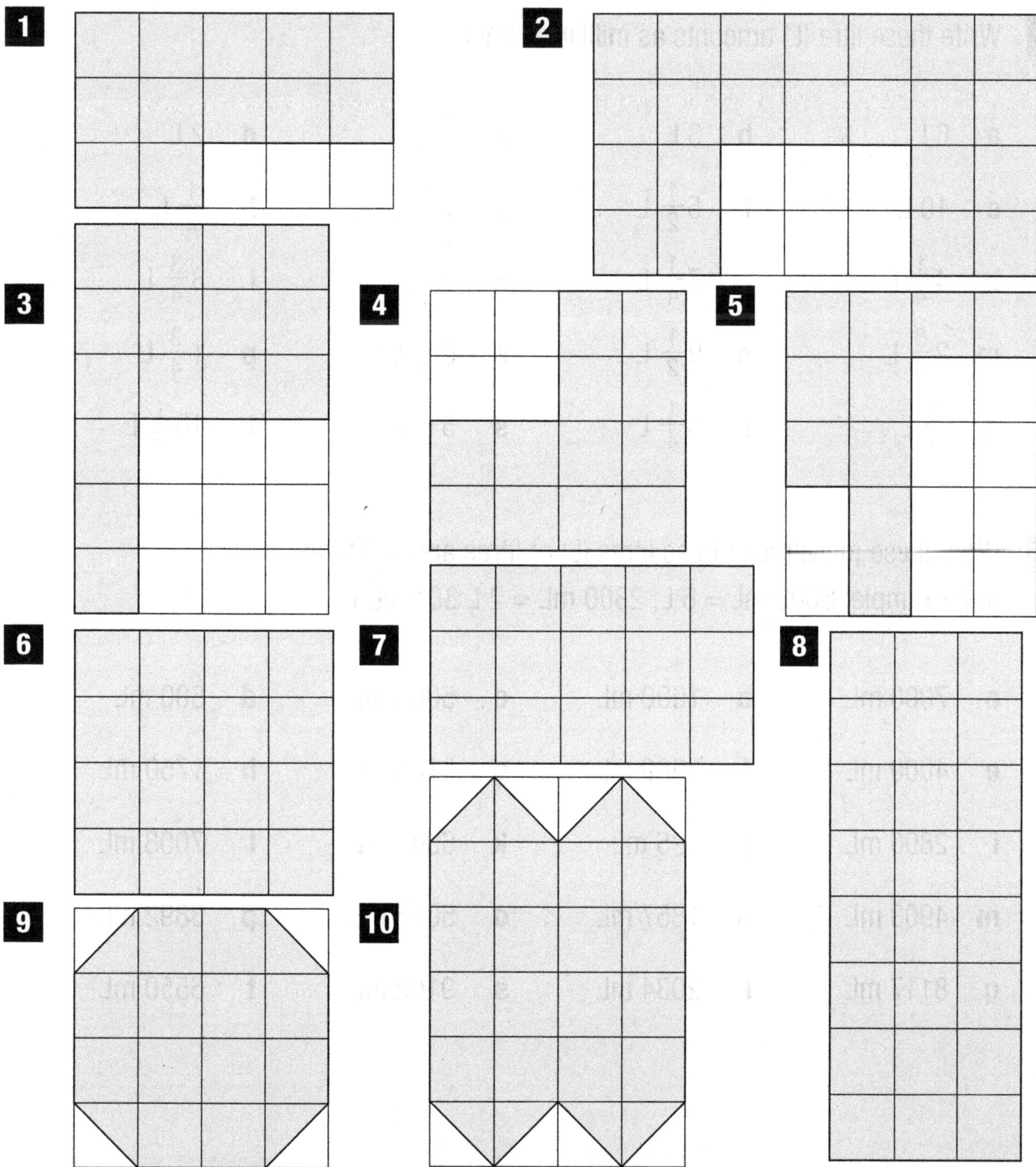

Estimate and measure volume and capacity using standard units

Convert volume and capacity units

1 Write these litre (L) amounts as millilitres (mL).

a	6 L	**b**	3 L	**c**	9 L	**d**	2 L
e	10 L	**f**	$5\frac{1}{2}$ L	**g**	$1\frac{1}{2}$ L	**h**	$\frac{1}{2}$ L
i	$4\frac{1}{4}$ L	**j**	$7\frac{1}{4}$ L	**k**	$\frac{3}{4}$ L	**l**	$8\frac{3}{4}$ L
m	$2\frac{3}{4}$ L	**n**	$9\frac{1}{2}$ L	**o**	$6\frac{1}{4}$ L	**p**	$1\frac{3}{4}$ L
q	$\frac{1}{4}$ L	**r**	$3\frac{1}{4}$ L	**s**	$5\frac{3}{4}$ L	**t**	$10\frac{1}{2}$ L

2 Write these millilitres (mL) as litres (L) or litres and millilitres.
(For example: 6000 mL = 6 L; 2300 mL = 2 L 300 mL.)

a	7000 mL	**b**	1000 mL	**c**	5000 mL	**d**	500 mL
e	4000 mL	**f**	8600 mL	**g**	3100 mL	**h**	1750 mL
i	2890 mL	**j**	795 mL	**k**	6345 mL	**l**	7008 mL
m	4903 mL	**n**	1857 mL	**o**	5076 mL	**p**	3892 mL
q	8117 mL	**r**	2034 mL	**s**	9102 mL	**t**	6550 mL

3 Write all the amounts that are more than 1 L.

1500 mL	$\frac{1}{2}$ L	800 mL	$2\frac{1}{4}$ L	1010 mL
710 mL	$5\frac{1}{2}$ L	150 mL	$\frac{3}{4}$ L	2500 mL

4 Write all the amounts that are more than 500 mL.

$\frac{3}{4}$ L	1 L	$\frac{1}{4}$ L	305 mL	$1\frac{1}{4}$ L
900 mL	110 mL	$2\frac{1}{2}$ L	450 mL	505 mL

5 Write all the amounts that are between 750 mL and $1\frac{1}{2}$ L.

1 L	$\frac{1}{4}$ L	900 mL	1200 mL	1750 mL
$1\frac{1}{4}$ L	1050 mL	$2\frac{1}{4}$ L	800 mL	$\frac{1}{2}$ L

6 Write all the amounts that are between $1\frac{3}{4}$ L and 2250 mL.

1500 mL	2 L	1800 mL	$2\frac{1}{2}$ L	$2\frac{3}{4}$ L
1790 mL	2030 mL	$1\frac{1}{4}$ L	2300 mL	3 L

Estimate and measure weight of objects using standard units

Convert weight units

1 Work out how many grams of each product are needed.

a $\frac{1}{2}$ kg dried fruit
b $\frac{1}{4}$ kg sugar
c $\frac{3}{4}$ kg bananas
d 1 kg plain flour
e $2\frac{1}{2}$ kg potatoes
f $1\frac{1}{4}$ kg tomatoes
g 2 kg onions
h $1\frac{1}{2}$ kg sausages
i $1\frac{3}{4}$ kg kaukau
j $\frac{1}{10}$ kg milk powder
k $\frac{9}{10}$ kg butter
l $\frac{3}{10}$ kg biscuits
m $\frac{7}{10}$ kg salt
n $1\frac{1}{10}$ kg rice
o $2\frac{3}{10}$ kg sago

2 Write these grams (g) as kilograms (kg) or kilograms and grams.
(For example: 3000 g = 3 kg; 1700 g = 1 kg 700 g)

a 8000 g
b 2000 g
c 1000 g
d 7000 g
e 4500 g
f 2800 g
g 3500 g
h 8100 g
i 850 g
j 1170 g
k 5980 g
l 6700 g
m 9500 g
n 8465 g
o 1014 g
p 2063 g
q 6005 g
r 5010 g
s 3609 g
t 1401 g

3 Write all the weights that are more than 1 kg.

700 g	1300 g	$\frac{1}{2}$ kg	$1\frac{1}{4}$ kg	915 g
650 g	$2\frac{1}{2}$ kg	$\frac{3}{4}$ kg	1050 g	1250 g

4 Write all the weights that are more than $\frac{1}{2}$ kg.

450 g	750 g	1000 g	110 g	$\frac{1}{4}$ kg
$1\frac{1}{2}$ kg	250 g	900 g	670 g	$\frac{3}{4}$ kg

5 Write all the weights that are between 700 g and $1\frac{1}{4}$ kg.

$\frac{1}{2}$ kg	1000 g	$\frac{3}{4}$ kg	1500 g	2000 g
1220 g	$\frac{1}{4}$ kg	$1\frac{1}{2}$ kg	650 g	1150 g

6 Write all the weights that are between 1400 g and $2\frac{1}{4}$ kg.

$1\frac{1}{2}$ kg	2000 g	2600 g	$1\frac{3}{4}$ kg	1970 g
1 kg	$1\frac{1}{4}$ kg	1320 g	2150 g	$2\frac{1}{2}$ kg

Tell and read time accurately to the quarter hour

Revise o'clock and half-past times

Write the time from each clock face in analogue and digital form. For example, the first clock shows 8 o'clock or 8:00.

1

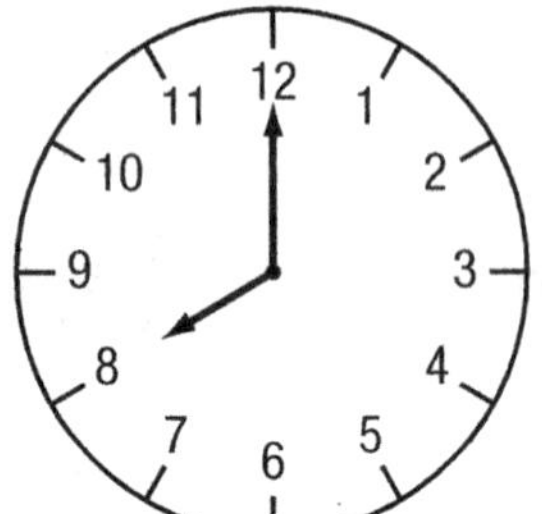

2

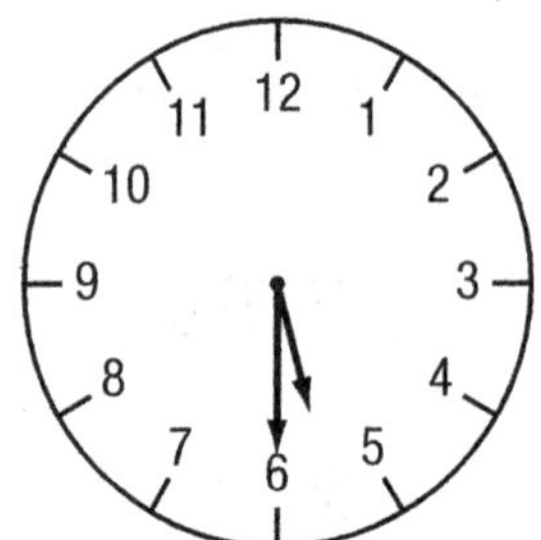

3

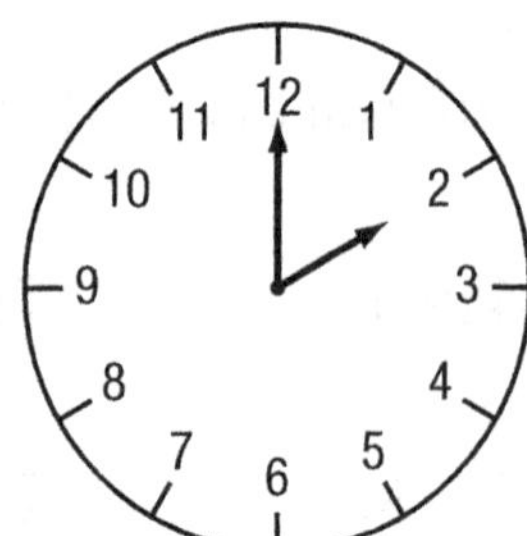

4

5

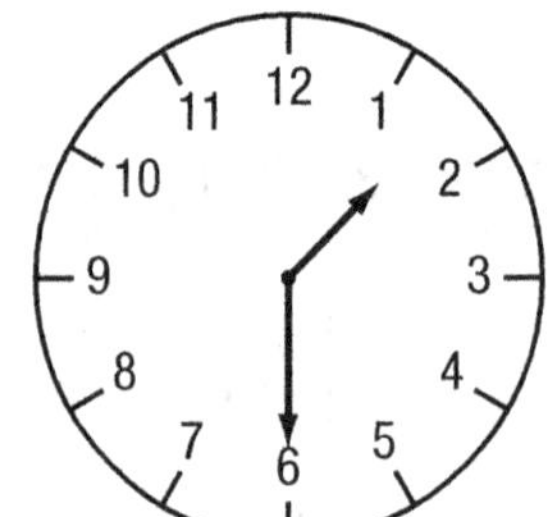

6

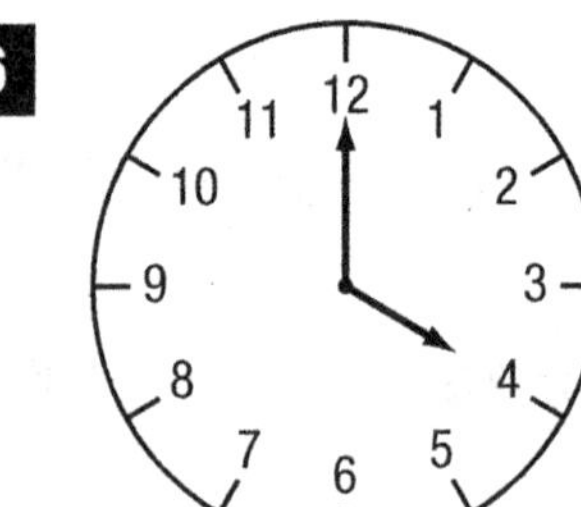

7

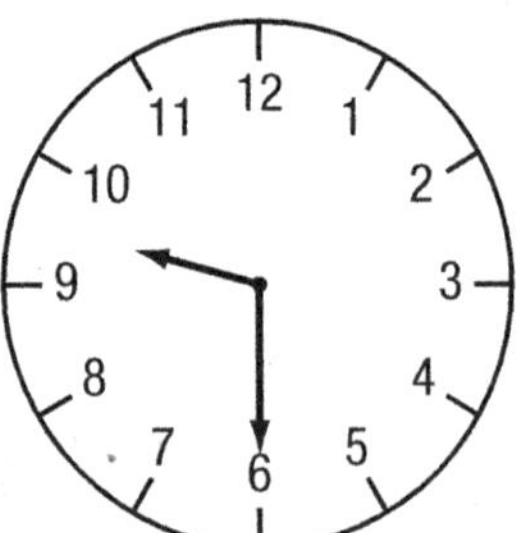

8

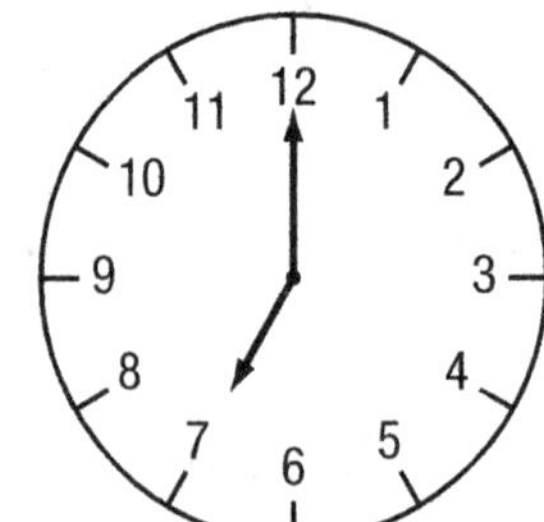

9

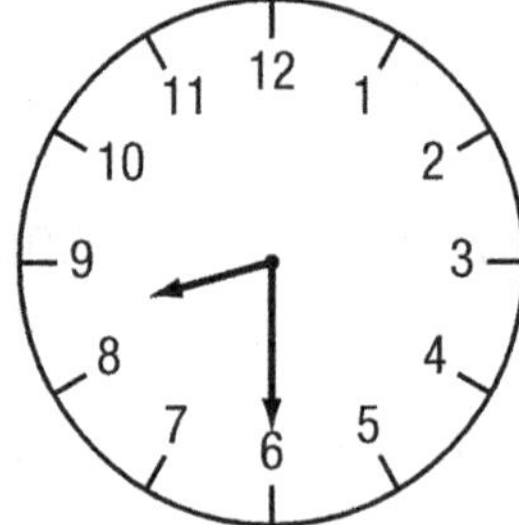

10

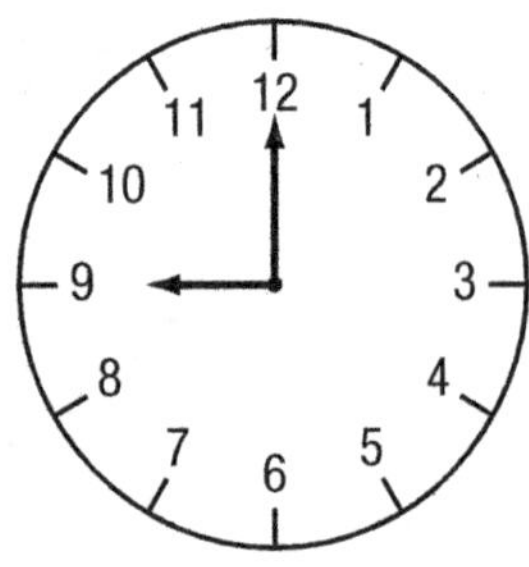

11

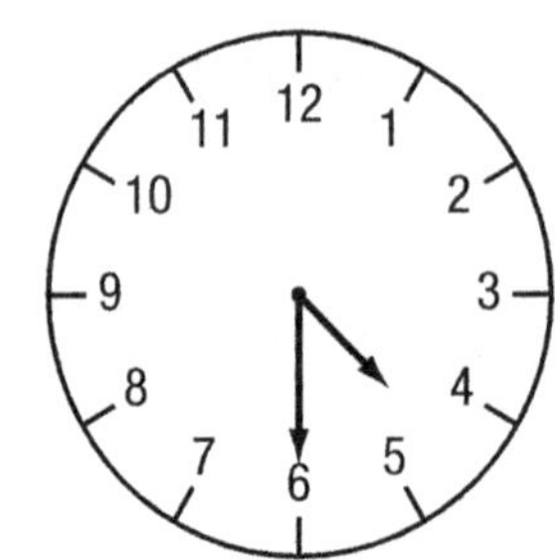

12

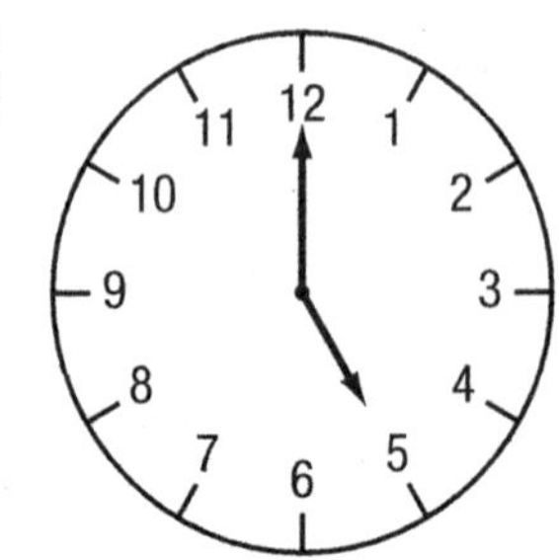

Read and record times to the quarter hour

1 Write the time from each clock face in analogue form. For example, the first clock shows quarter past 7.

a 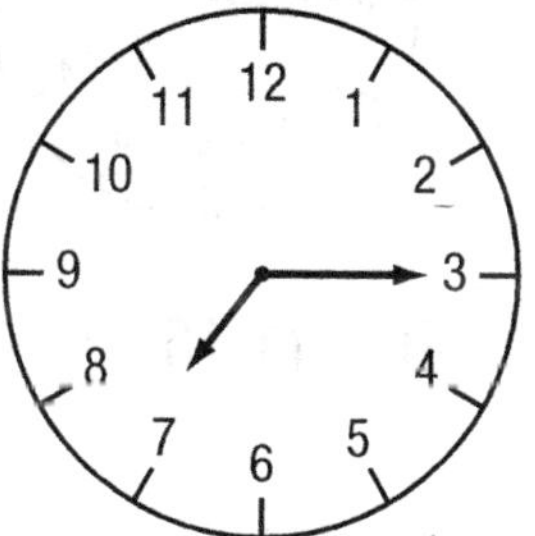

b 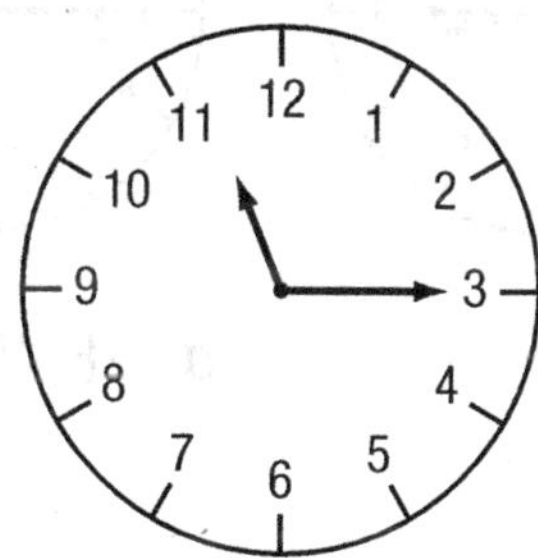

c 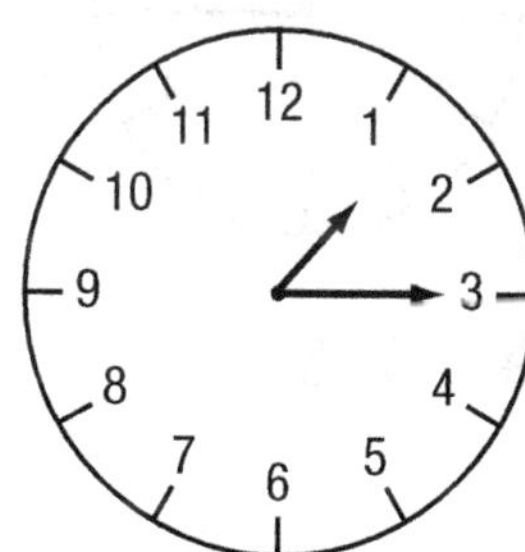

d 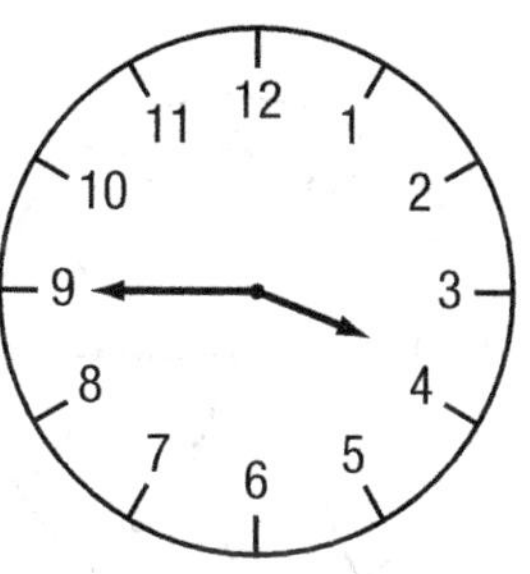

e 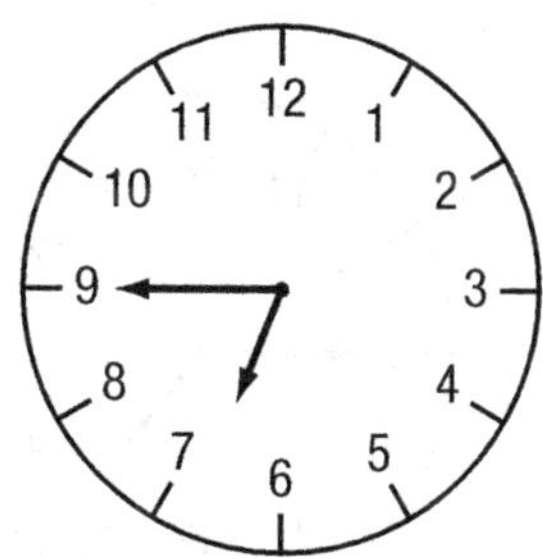

f 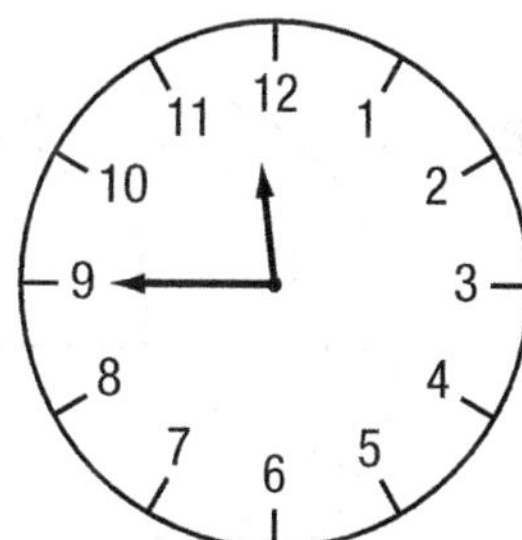

g 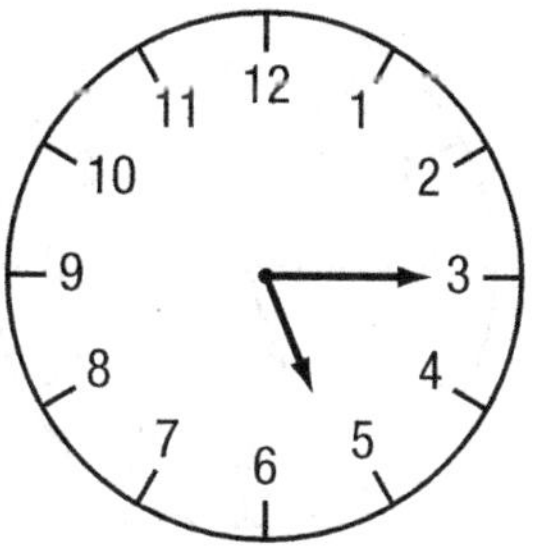

h 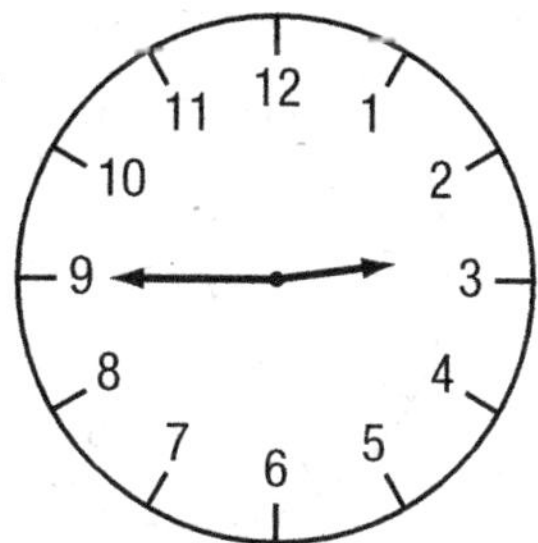

i 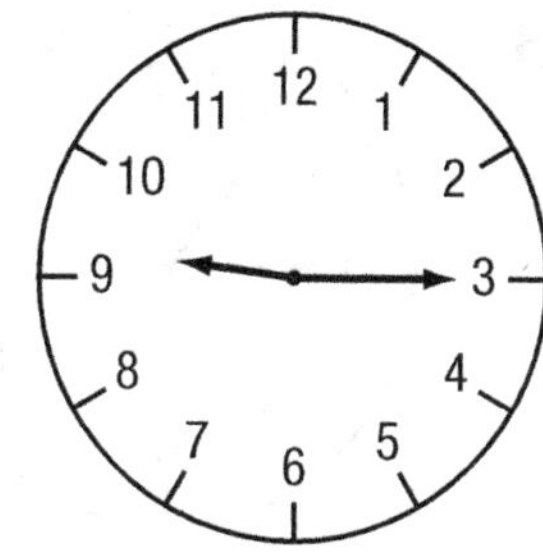

j 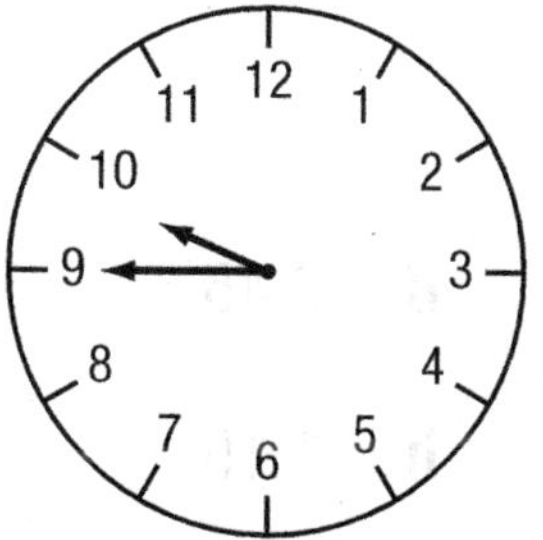

k 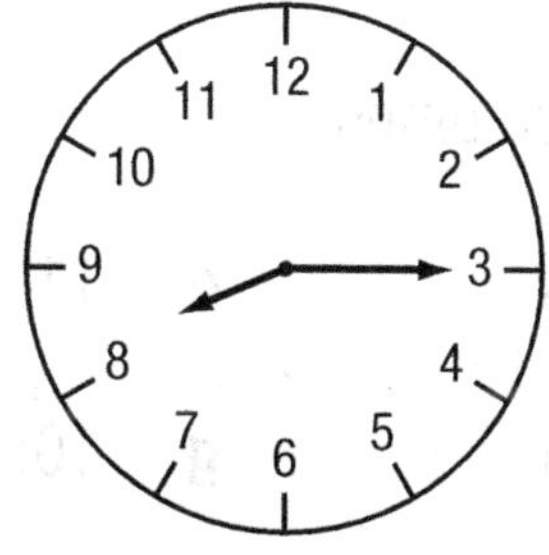

l 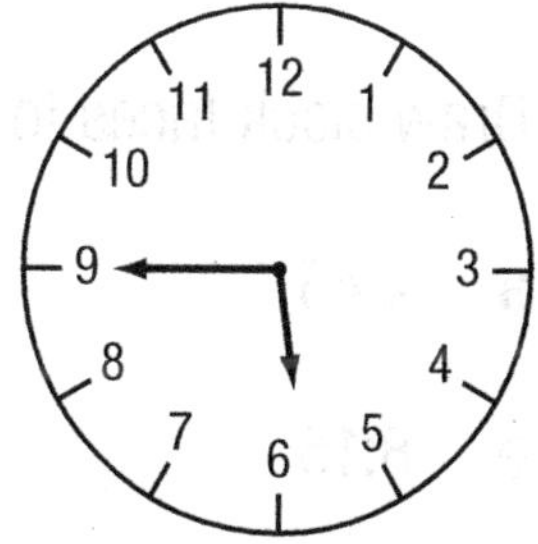

2 Match the analogue and digital clock times.

a

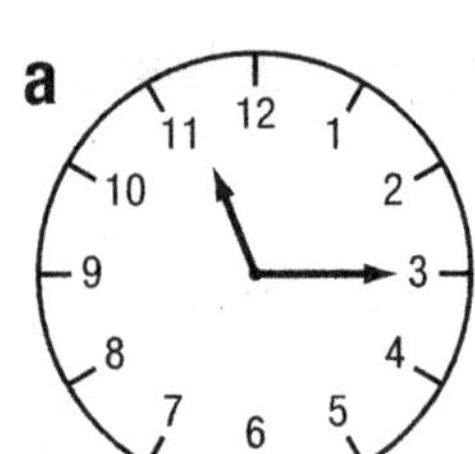

b

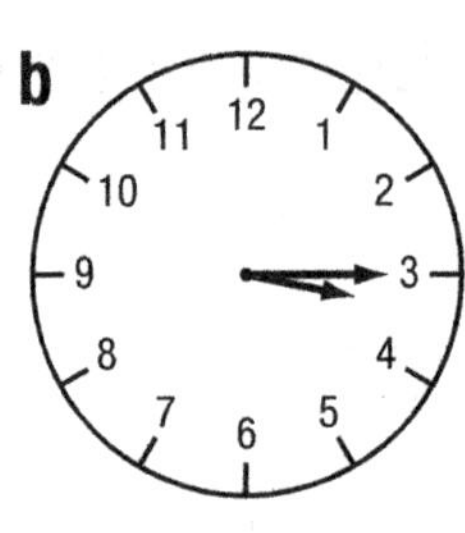

c

d

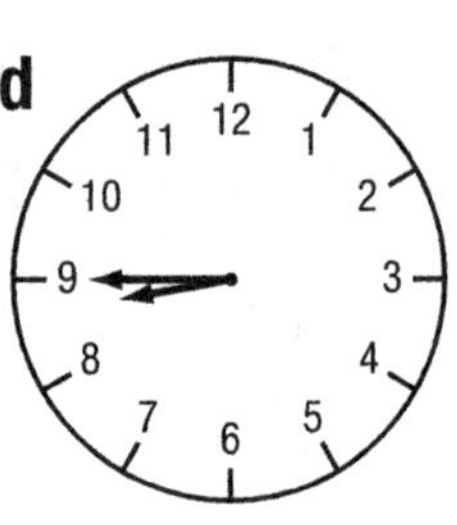

e 3:15 **f** 5:45 **g** 8:45 **h** 11:15

3 Write the time from each clock face in analogue and digital form.

a

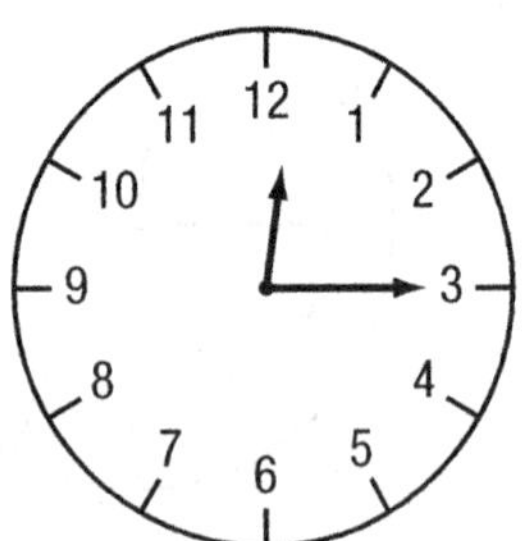

b

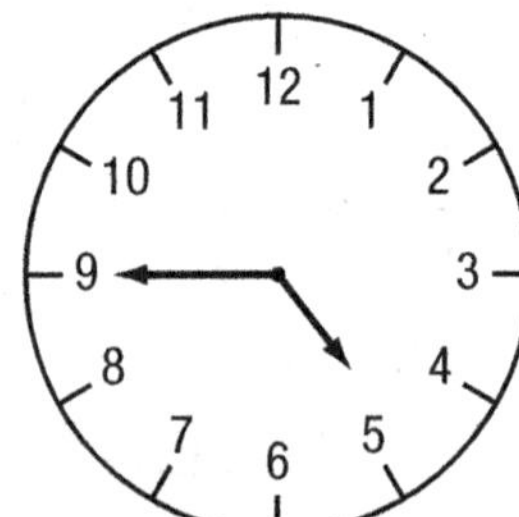

c

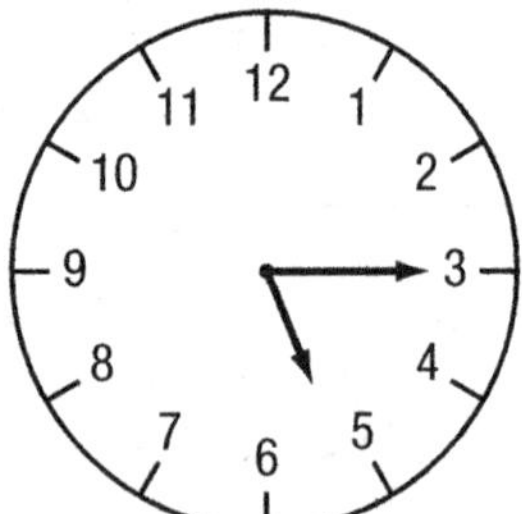

d

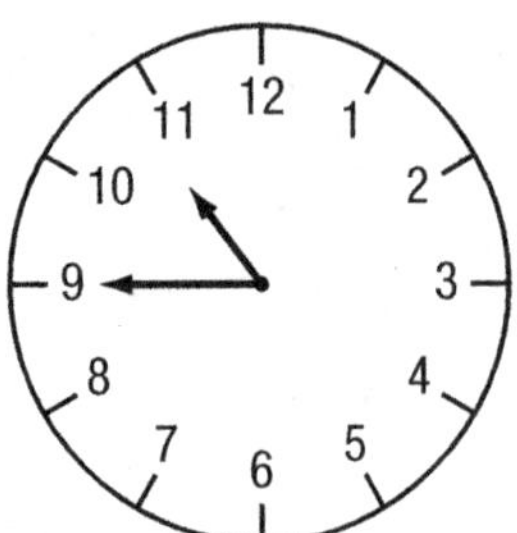

e

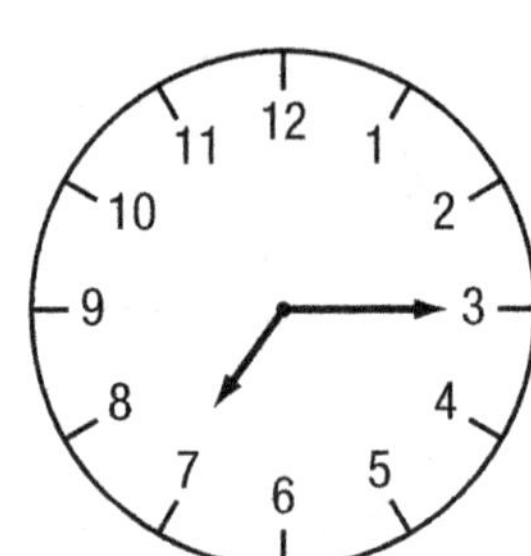

f

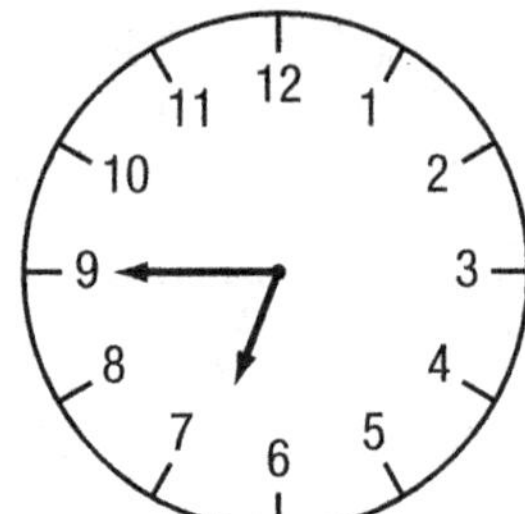

4 Draw clock faces to show these times.

a 2:45 **b** 6:15 **c** 11:45 **d** 3:45

e 8:15 **f** 1:45 **g** 10:15 **h** 1:15

Convert between time units

1 How many months in these years?

a	1 year	**b**	5 years	**c**	10 years	**d**	4 years	**e**	8 years
f	2 years	**g**	7 years	**h**	3 years	**i**	6 years	**j**	$1\frac{1}{2}$ years
k	$4\frac{1}{2}$ years	**l**	$2\frac{1}{2}$ years	**m**	$7\frac{1}{2}$ years	**n**	$5\frac{1}{2}$ years	**o**	$9\frac{1}{2}$ years

2 How many days in these weeks?

a	2 weeks	**b**	10 weeks	**c**	4 weeks	**d**	5 weeks
e	8 weeks	**f**	1 week	**g**	3 weeks	**h**	9 weeks
i	11 weeks	**j**	7 weeks	**k**	12 weeks	**l**	6 weeks

3 How many minutes in these hours?

a	1 hour	**b**	5 hours	**c**	2 hours	**d**	10 hours	**e**	3 hours
f	8 hours	**g**	6 hours	**h**	9 hours	**i**	7 hours	**j**	$1\frac{1}{2}$ hours
k	$5\frac{1}{2}$ hours	**l**	$3\frac{1}{2}$ hours	**m**	$4\frac{1}{2}$ hours	**n**	$2\frac{1}{2}$ hours	**o**	$8\frac{1}{2}$ hours

4 How many hours in these days?

a	2 days	**b**	4 days	**c**	1 day	**d**	8 days	**e**	5 days
f	10 days	**g**	3 days	**h**	7 days	**i**	6 days	**j**	$1\frac{1}{2}$ days
k	$5\frac{1}{2}$ days	**l**	$9\frac{1}{2}$ days	**m**	$3\frac{1}{2}$ days	**n**	$6\frac{1}{2}$ days	**o**	$2\frac{1}{2}$ days

Calculate time before and after a given time

Look at each clock and work out the time.

1

a 1 hour later

b 1 hour earlier

c 2 hours earlier

d $\frac{1}{2}$ hour later

2 4:15

a $\frac{1}{2}$ hour earlier

b $\frac{1}{4}$ hour later

c 1 hour later

d 3 hours earlier

3

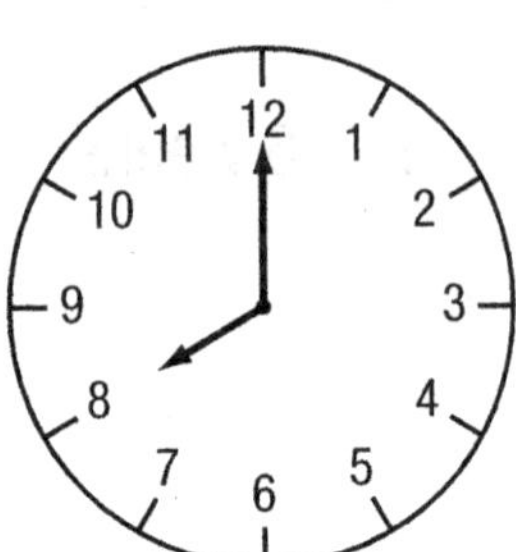

a 15 minutes earlier

b 1 hour earlier

c $\frac{1}{2}$ hour later

d $\frac{1}{4}$ hour later

4 10:45

a $\frac{1}{2}$ hour later

b $\frac{1}{4}$ hour earlier

c 1 hour later

d 15 minutes later

5

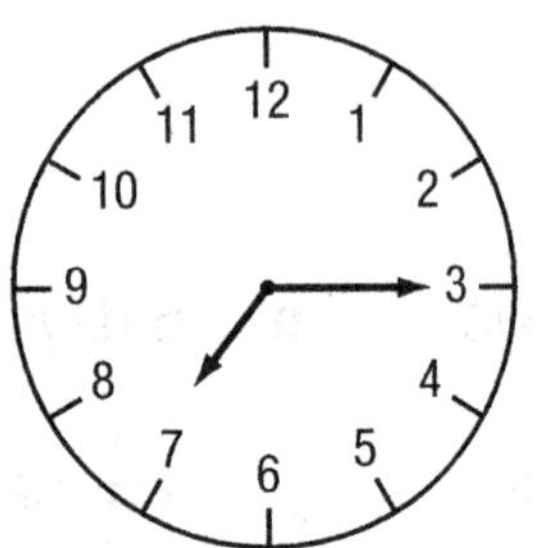

a 15 minutes later

b 15 minutes earlier

c $1\frac{1}{2}$ hours later

d $\frac{1}{2}$ hour earlier

6 9:30

a 1 hour earlier **b** $1\frac{1}{4}$ hours earlier

c 30 minutes later **d** $\frac{1}{4}$ hour later

7

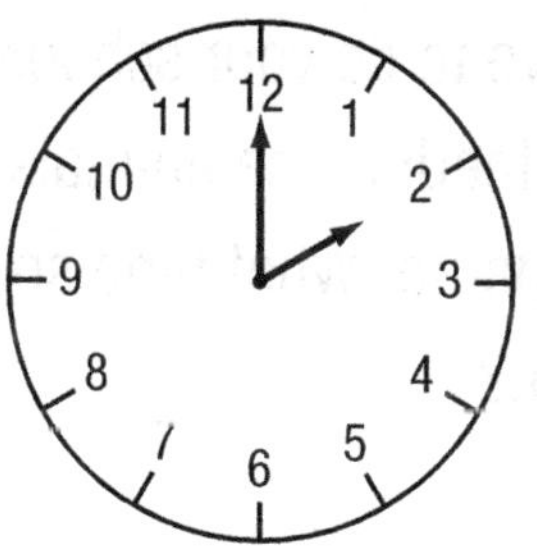

a $\frac{1}{2}$ hour earlier **b** 15 minutes later

c $1\frac{1}{4}$ hours later **d** $\frac{1}{4}$ hour earlier

8 2:45

a $\frac{1}{4}$ hour later **b** $1\frac{1}{2}$ hours earlier

c $1\frac{1}{2}$ hours later **d** 15 minutes earlier

Solve time problems

1 You have to be at work at 6:30. It will take you one hour to get there. What time should you leave home?

2 You want to eat dinner at 7 o'clock. It will take $1\frac{1}{2}$ hours to prepare. What time should you start preparing?

3 The bus leaves at 3:30. You live $\frac{1}{4}$ hour from the bus stop. What time should you leave home?

4 Breakfast is at 7:15. It takes 15 minutes to have a shower and get dressed. What time should you set the alarm to wake you up?

5 The soccer game starts at 5 o'clock. Before you go, you have to do your schoolwork and feed the animals. Schoolwork will take $\frac{1}{2}$ an hour and feeding the animals will take 15 minutes. It will take you 30 minutes to walk to the game. What time should you start these tasks so you can be at the game at 5 o'clock?

6 Our plane leaves at 7:45. We have to be at the airport 2 hours before and it will take us $\frac{3}{4}$ of an hour to get there. What time should we leave home?

7 The concert starts at 1:15. It takes $\frac{1}{2}$ an hour to get there and 15 minutes to park the car and buy tickets. What time should you leave home?

8 Your favourite TV show starts at 5:30. You have to water the garden for $\frac{1}{4}$ of an hour. You also have to peel and cut up vegetables for dinner, which should take you $\frac{1}{2}$ an hour. What time should you start these tasks so you can watch TV?

Assessment Measurement

1 Measure each line and answer the questions below.

A

B

C

D

E

Which line has a length of:

a 60 mm? **b** 45 mm? **c** 11 cm? **d** $7\frac{1}{2}$ cm?

e Order the lines from longest to shortest.

2 Write these lengths as centimetres.

a 7 m **b** 2.5 m **c** $4\frac{1}{2}$ m **d** 3.9 m

e 0.7 m **f** 1.63 m **g** $2\frac{1}{2}$ m **h** 5.88 m

3 Write these lengths as metres.

a 800 cm **b** 550 cm **c** 1400 cm **d** 1850 cm

e 720 cm **f** 290 cm **g** 40 cm **h** 113 cm

4 Write these lengths as millimetres.

a 7 cm	**b** 2 cm	**c** $3\frac{1}{2}$ cm	**d** 6.5 cm
e 0.8 cm	**f** 8.1 cm	**g** 4.7 cm	**h** 1.3 cm

5 Write these lengths as centimetres.

a 30 mm	**b** 110 mm	**c** 75 mm	**d** 255 mm
e 400 mm	**f** 318 mm	**g** 506 mm	**h** 57 mm

6 Measure the perimeter of each shape in centimetres.

a

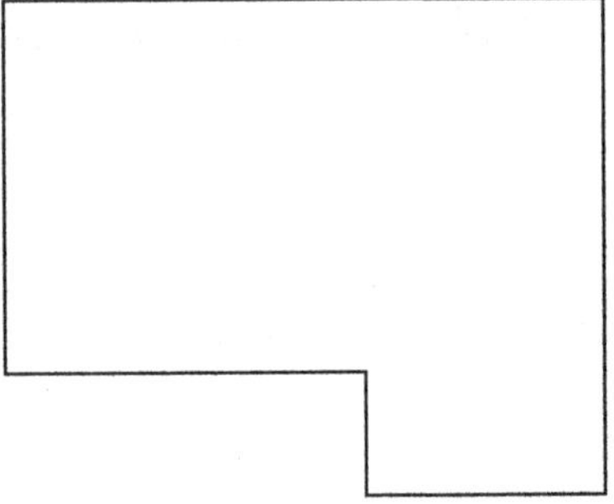

b

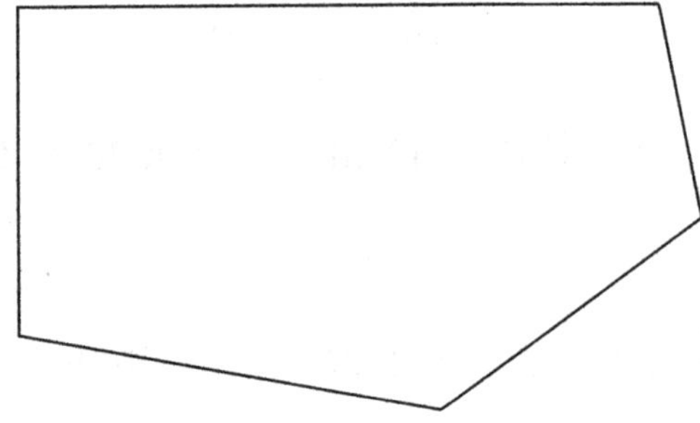

7 If each small square is 1 square centimetre, what is the area of each shape?

a

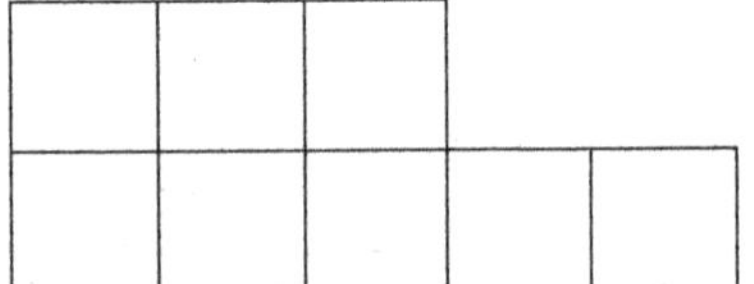

b

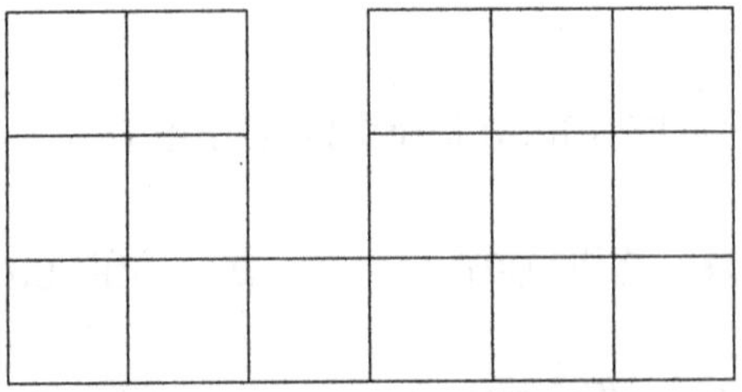

8 Complete the gaps.

a $5 \text{ L} = \_\_\_\_\_ \text{ mL}$ b $2\frac{1}{2} \text{ L} = \_\_\_\_\_ \text{ mL}$ c $\frac{1}{4} \text{ L} = \_\_\_\_\_ \text{ mL}$

d $9000 \text{ mL} = \_\_\_\_\_ \text{ L}$ e $1500 \text{ mL} = \_\_\_\_\_ \text{ L}$ f $6000 \text{ mL} = \_\_\_\_\_ \text{ L}$

9 Which of these amounts are between 850 mL and 1300 mL?

a 1 L b $1\frac{1}{2}$ L c $\frac{3}{4}$ L d 910 mL

e $1\frac{1}{4}$ L f 2 L g $\frac{1}{2}$ L h $\frac{1}{4}$ L

10 Complete the gaps.

a $1 \text{ kg} = \_\_\_\_\_ \text{ g}$ b $\frac{1}{2} \text{ kg} = \_\_\_\_\_ \text{ g}$ c $3\frac{1}{4} \text{ kg} = \_\_\_\_\_ \text{ g}$

d $3000 \text{ g} = \_\_\_\_\_ \text{ kg}$ e $7500 \text{ g} = \_\_\_\_\_ \text{ kg}$ f $5000 \text{ g} = \_\_\_\_\_ \text{ kg}$

11 Which of these weights are less than $1\frac{1}{2}$ kg?

a 1000 g b 2000 g c 1100 g d 750 g

e 1600 g f 1250 g g 900 g h 1800 g

12 Write the time from each clock in analogue and digital form.

a 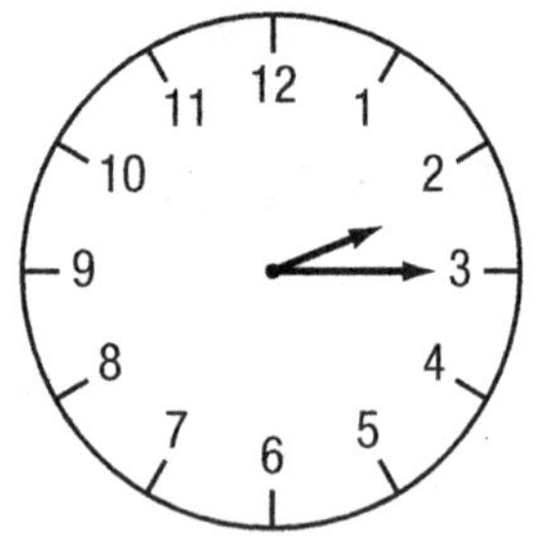

b 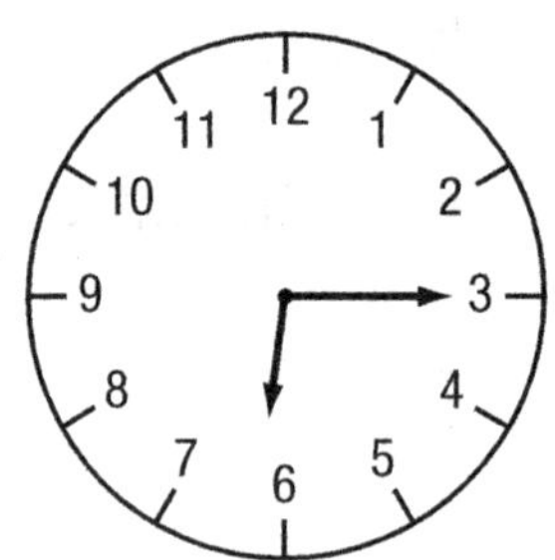

c 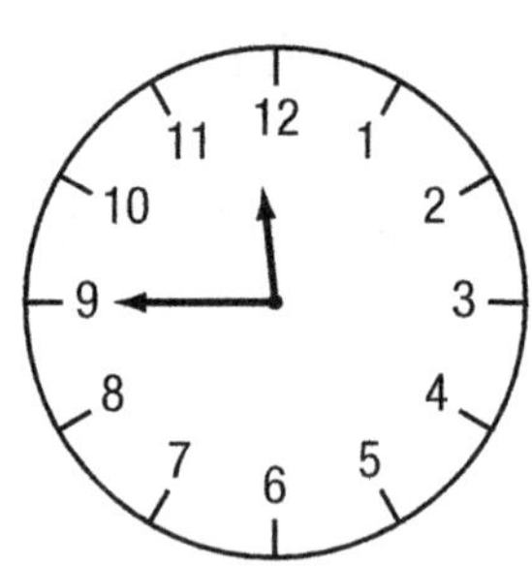

13 Draw clock faces to show these times.

a 3:30 **b** 1:15 **c** 9:45 **d** 7:45

14 Complete the gaps.

a 3 years = ______ months

b $3\frac{1}{2}$ years = ______ months

c 6 weeks = ______ days

d 7 weeks = ______ days

e 4 hours = ______ minutes

f $6\frac{1}{2}$ hours = ______ minutes

g 3 days = ______ hours

h $4\frac{1}{2}$ days = ______ hours

15 Work out the time:

2:30

a 1 hour later

b $\frac{1}{2}$ hour earlier

c 15 minutes later

d $1\frac{1}{4}$ hours earlier

16 Work out the time:

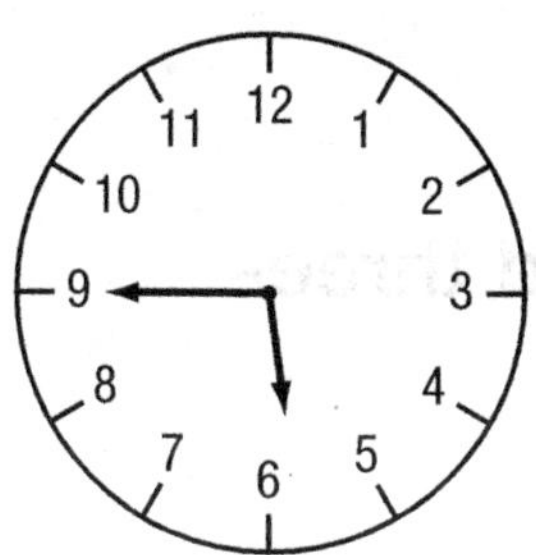

a $\frac{1}{2}$ hour later

b $\frac{1}{4}$ hour earlier

c $\frac{3}{4}$ hour later

d $1\frac{1}{2}$ hours earlier

17 Solve these word problems.

a Lunch has to be ready at 12:30. If it takes 45 minutes to prepare, what time should you start the preparation?

b You live 30 minutes from work and have to be there at 9:30. If it takes you 45 minutes to have a shower, get dressed and have breakfast, what time should you set your alarm?

Strand **Space and Shape**

Investigate and describe features of two and three-dimensional shapes

Recognise and name two and three-dimensional shapes

Write the name of each shape. Some shapes appear more than once.

1 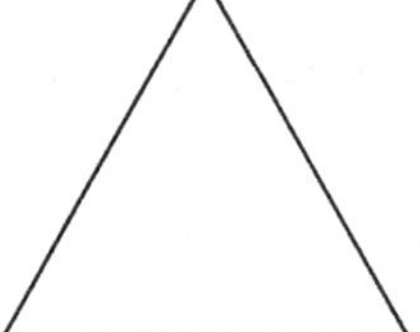

2 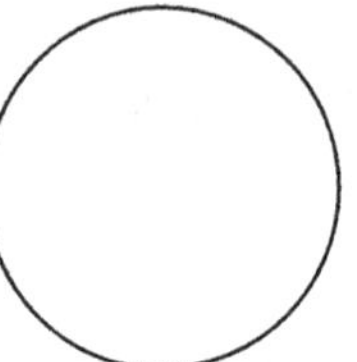

3 

4 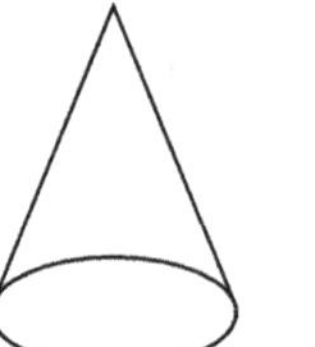

5 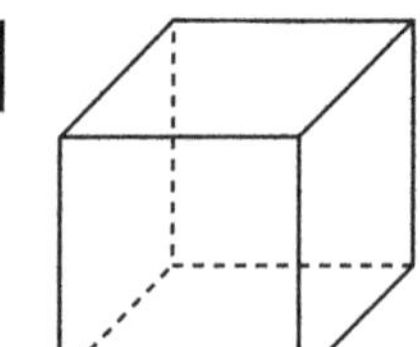

6 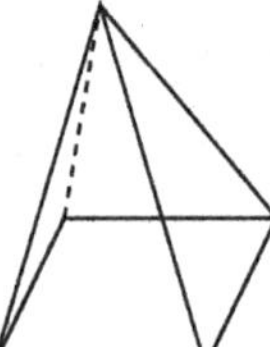

7 

8 

9 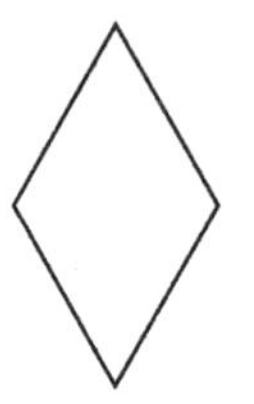

10 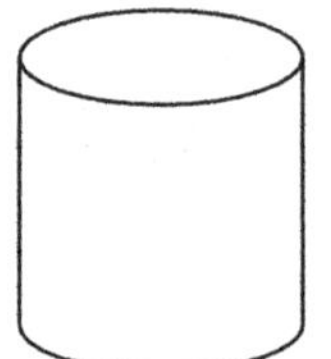

11 

12 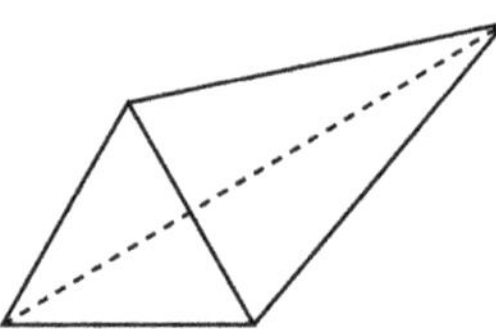

13 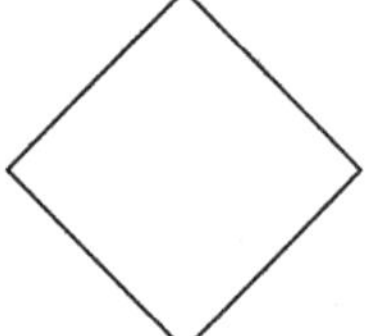

14 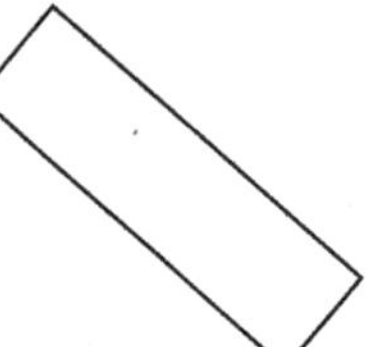

15 

Recognise features of two-dimensional shapes

Which shape in each group does not belong?

1

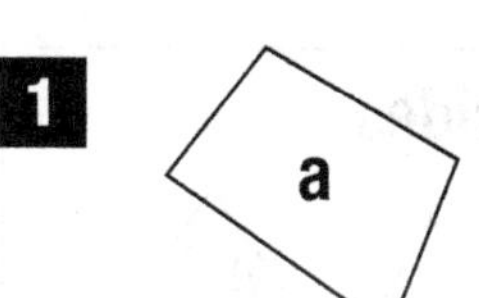

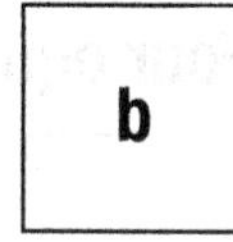

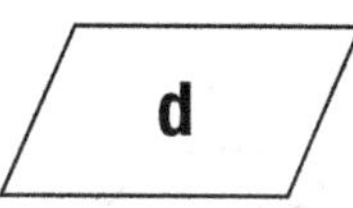

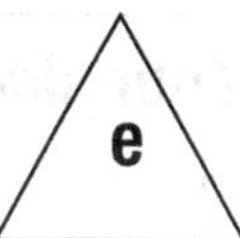

2

a
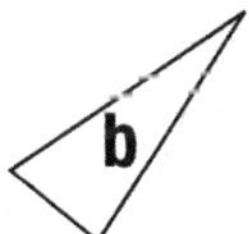

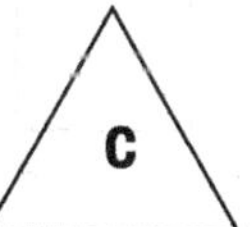

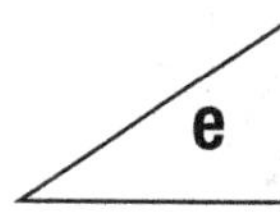

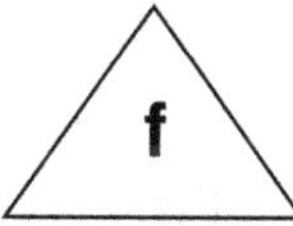

3

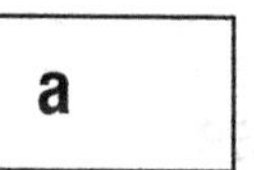

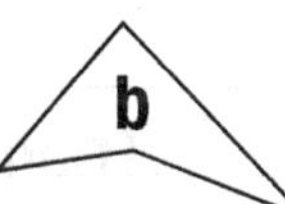

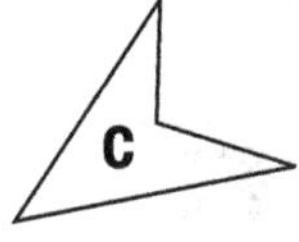

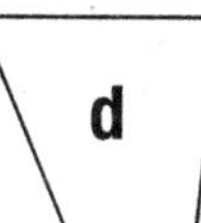

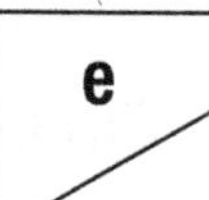

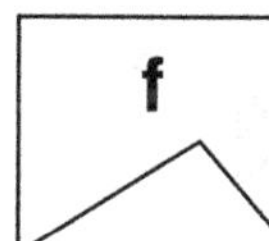

4

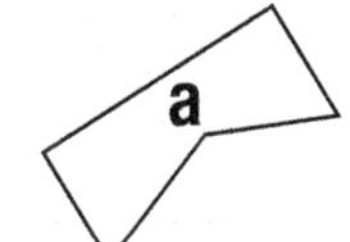

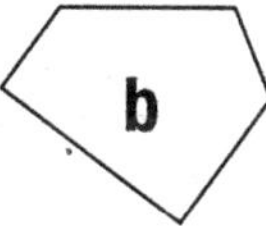

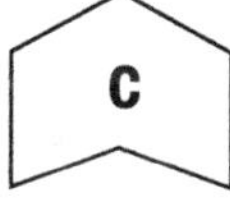

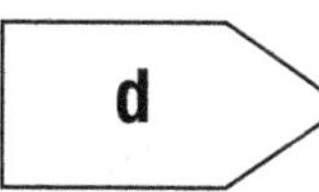

e
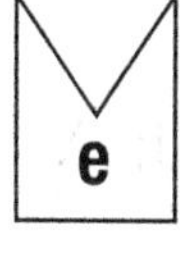

5

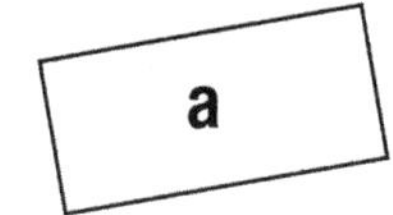

b
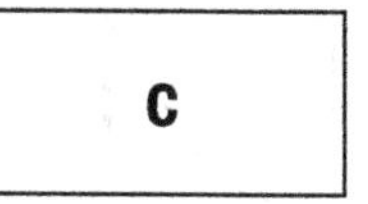

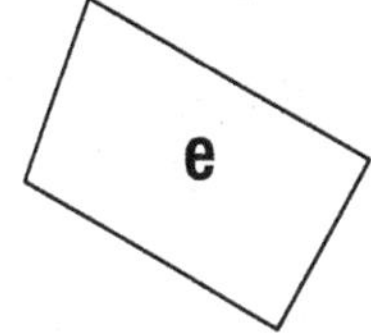

6

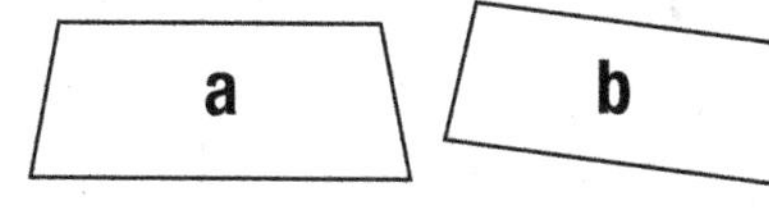

b

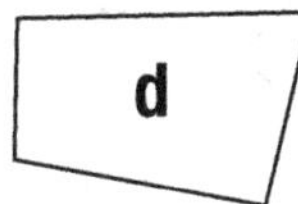

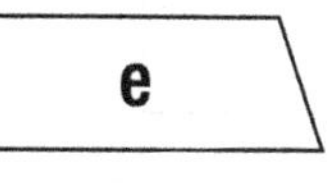

7

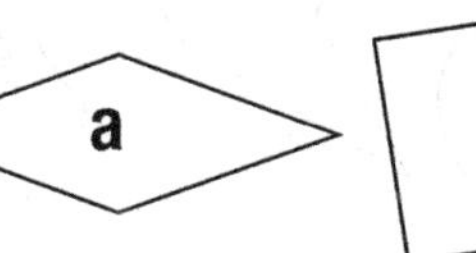

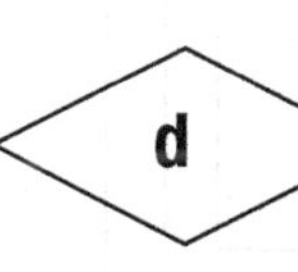

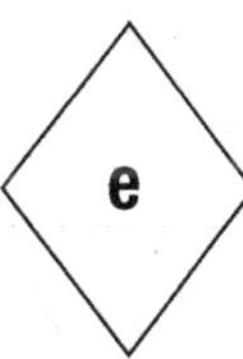

f

Identify, compare and classify two-dimensional shapes

Draw the shapes that fit in each group.

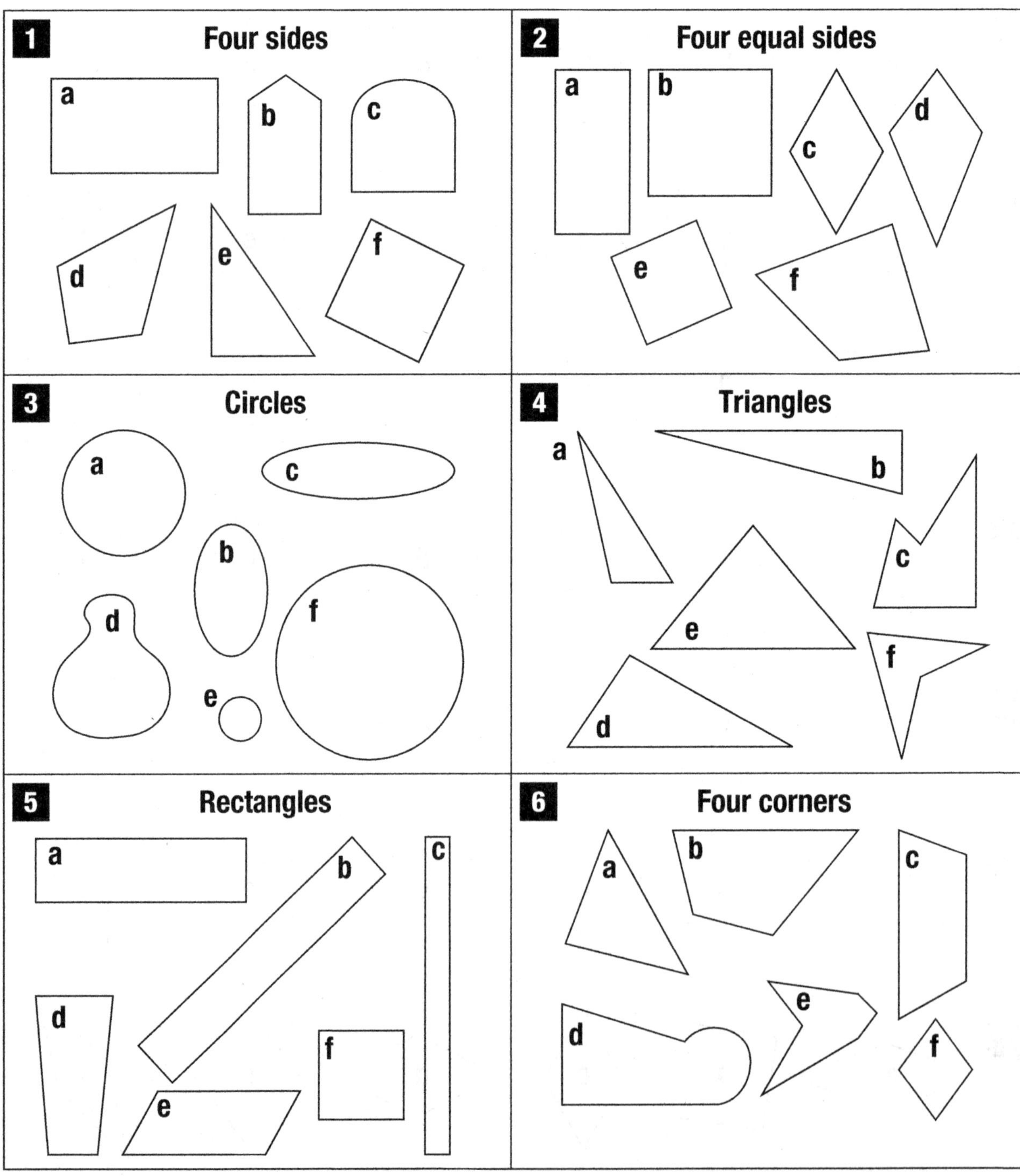

Identify, compare and classify three-dimensional shapes

Draw the shapes that fit in each group.

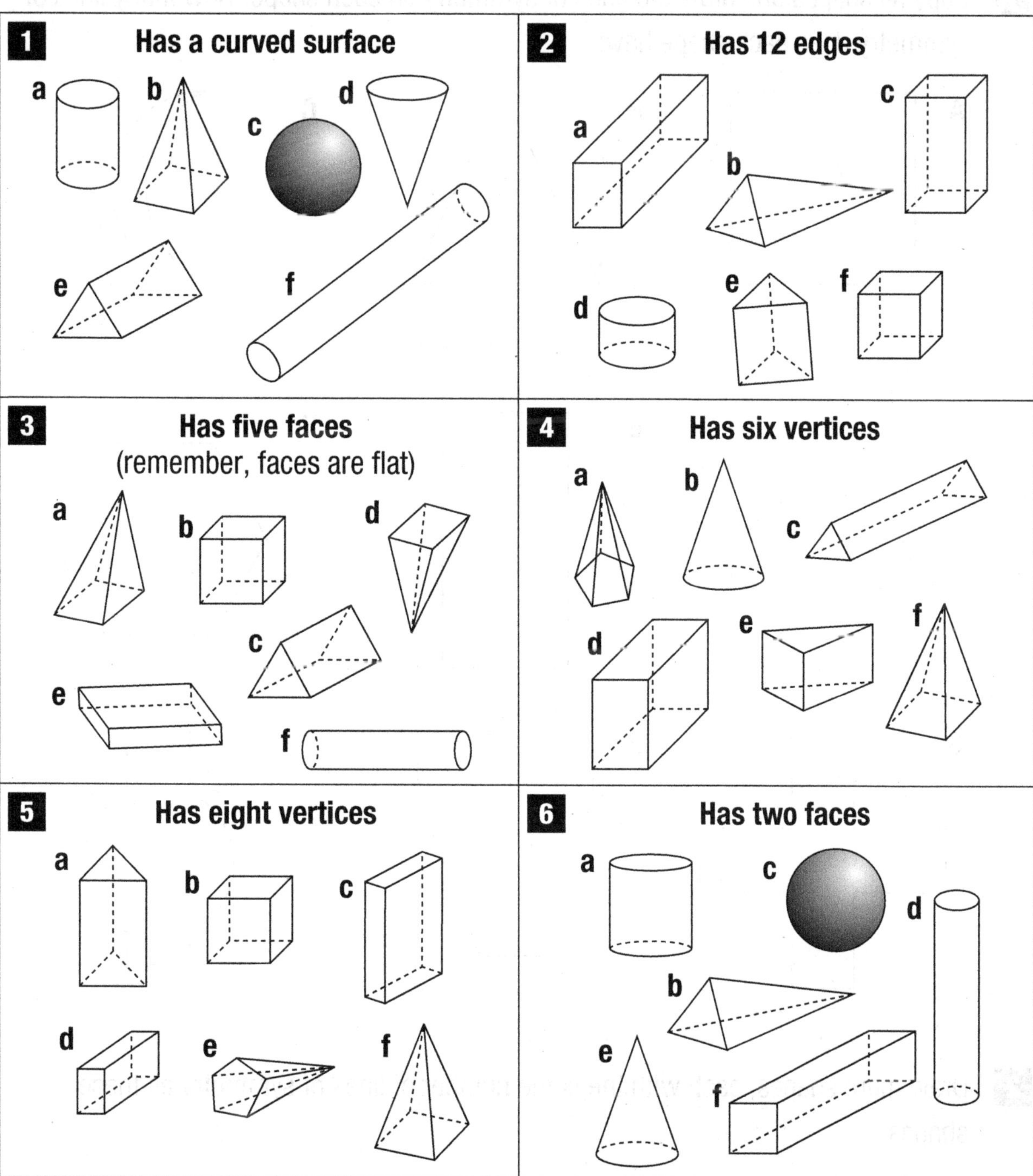

Identify lines of symmetry

1 Copy all shapes and draw the lines of symmetry on each shape. How many lines of symmetry does each shape have?

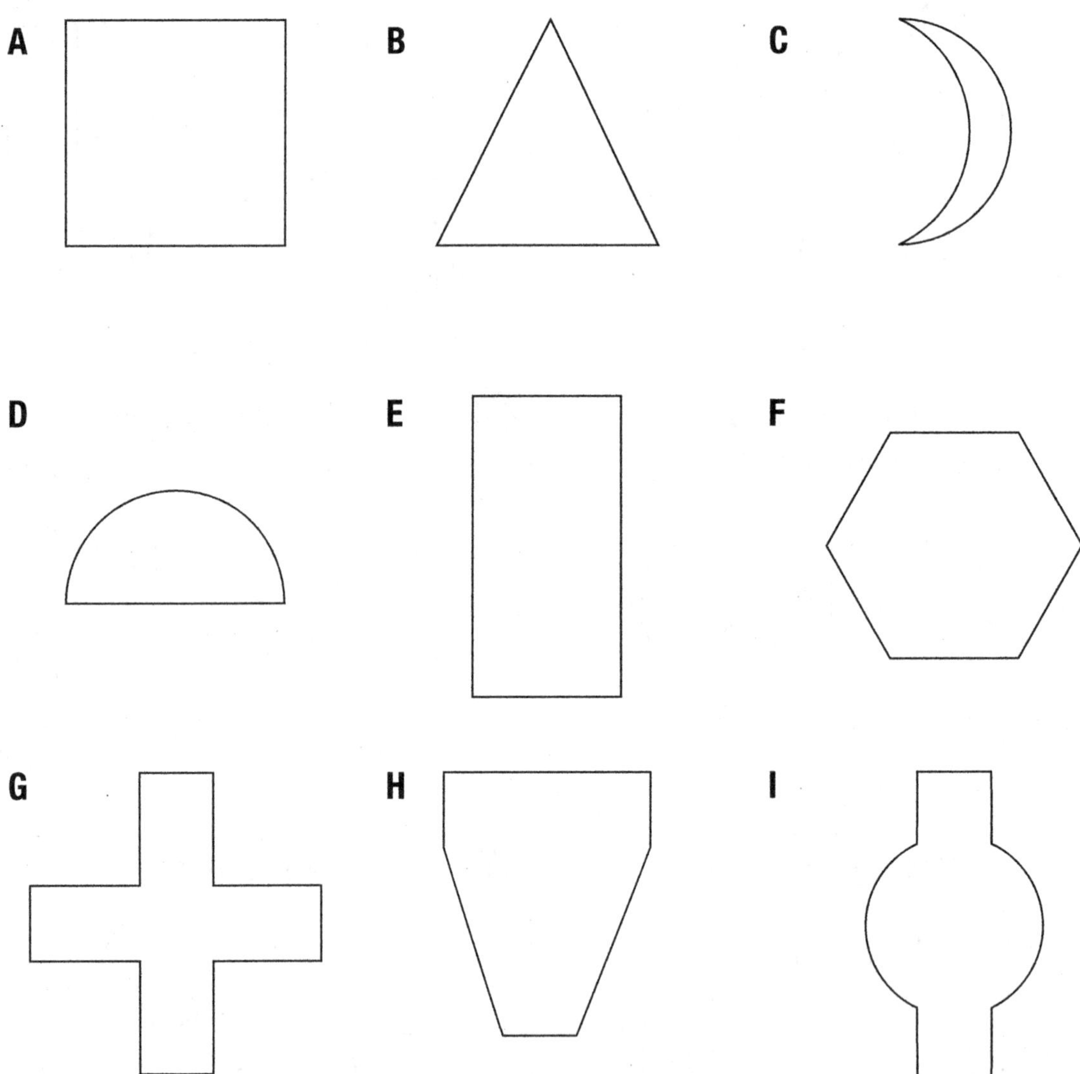

2 Draw more shapes, each with the same number of lines of symmetry as these shapes.

Recognise, name and describe angles

Identify angles

1 Find the 'quarter turn' angles.

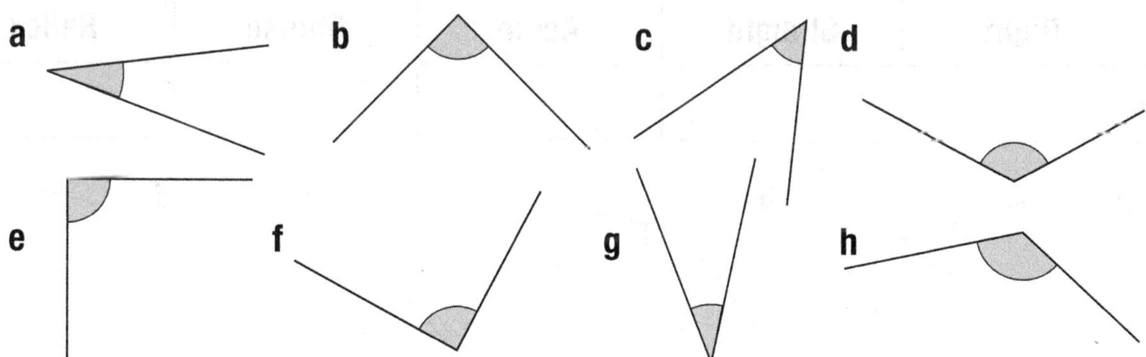

2 Find the angles that are more than a 'quarter turn'.

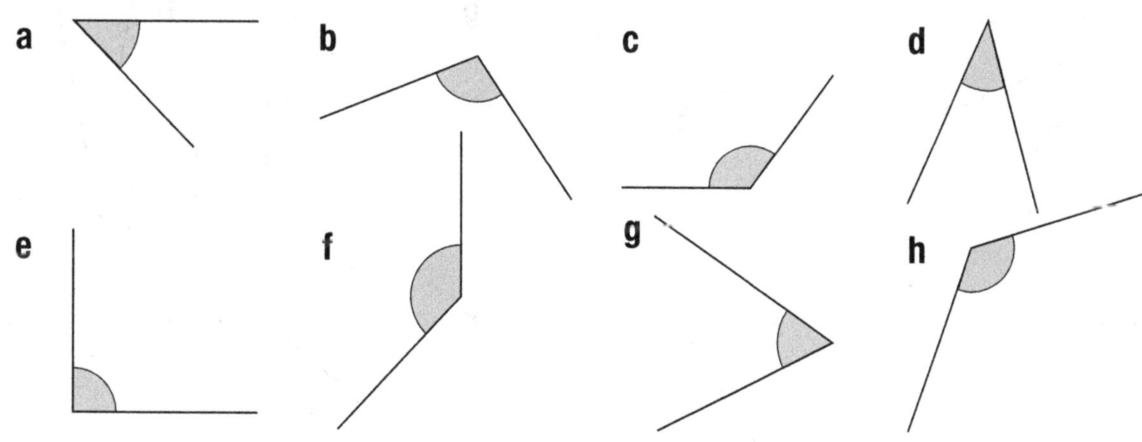

3 Find the angles that are less than a 'quarter turn'.

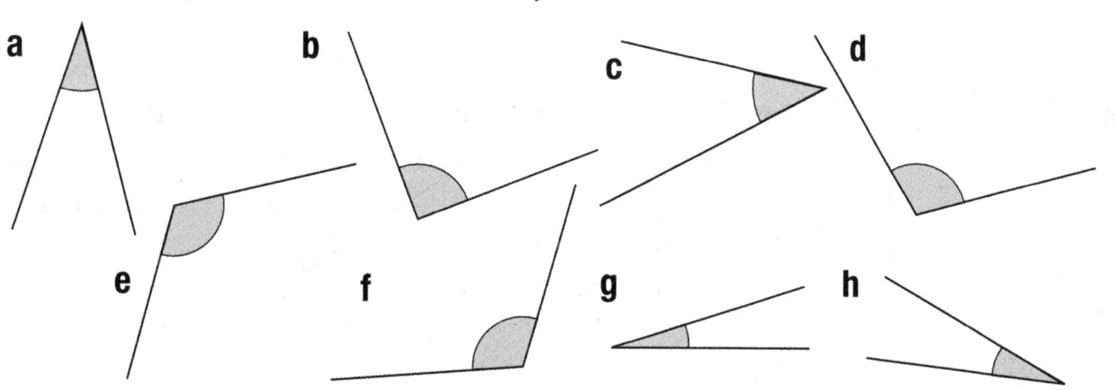

Recognise and name angles

Copy the following chart and write the letter for each angle in the correct place on the chart.

Right	Straight	Acute	Obtuse	Reflex

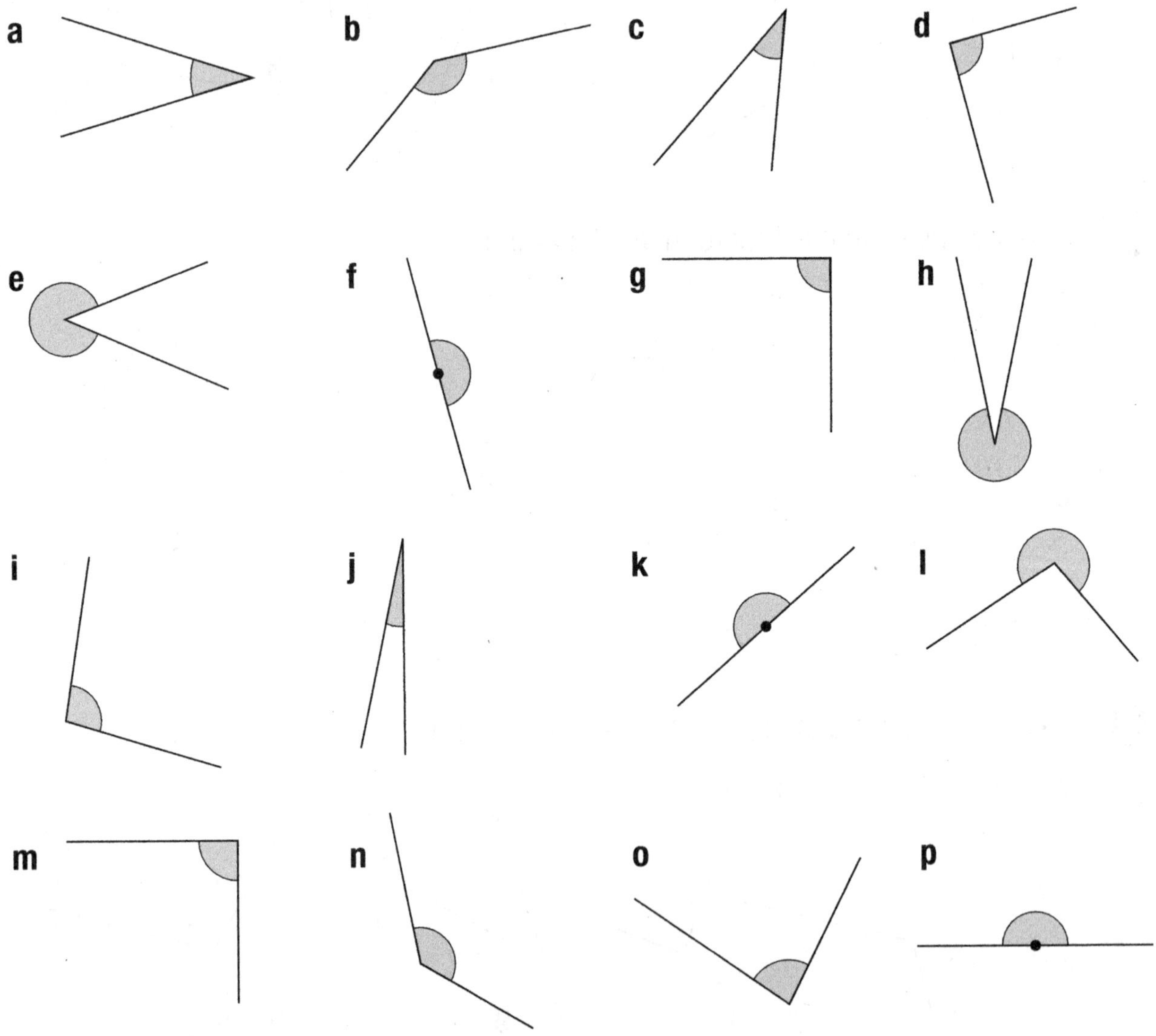

a

b

c

d

e

f

g

h

i

j

k

l

m

n

o

p

Compare and order angles

For each question, write the letters of the angles from smallest to largest.

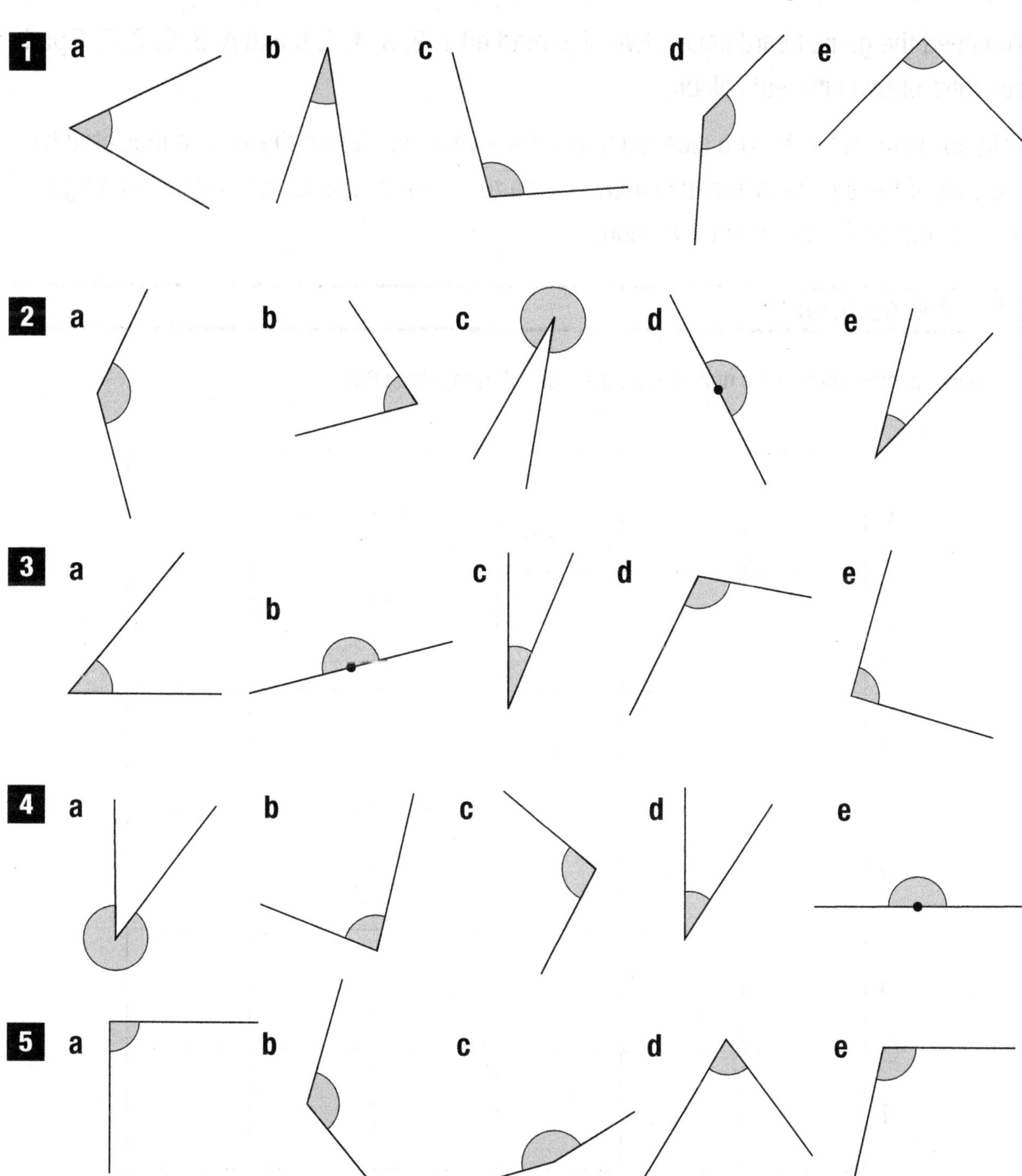

Describe positions on a grid

Play this game with a friend.

You need the game board below, two dice marked 1, 2, 3, 4, 5, 6 and A, B, C, D, E, F and counters of two different colours.

Take turns to throw the two dice together. Place your counter on the square indicated by the dice. If the square is already taken, miss a turn. The winner is the first person to get four counters in a row in any direction.

Remember

Read the letters before the numbers on the dice: C6 rather than 6C.

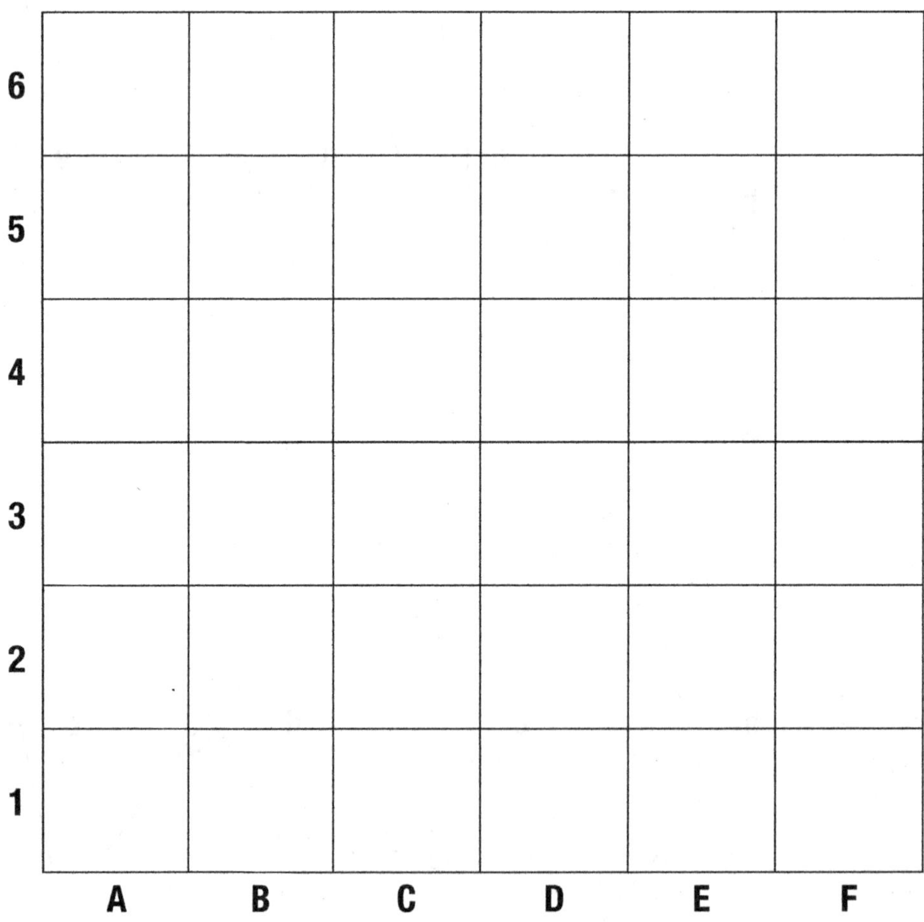

Give and describe positions on a grid

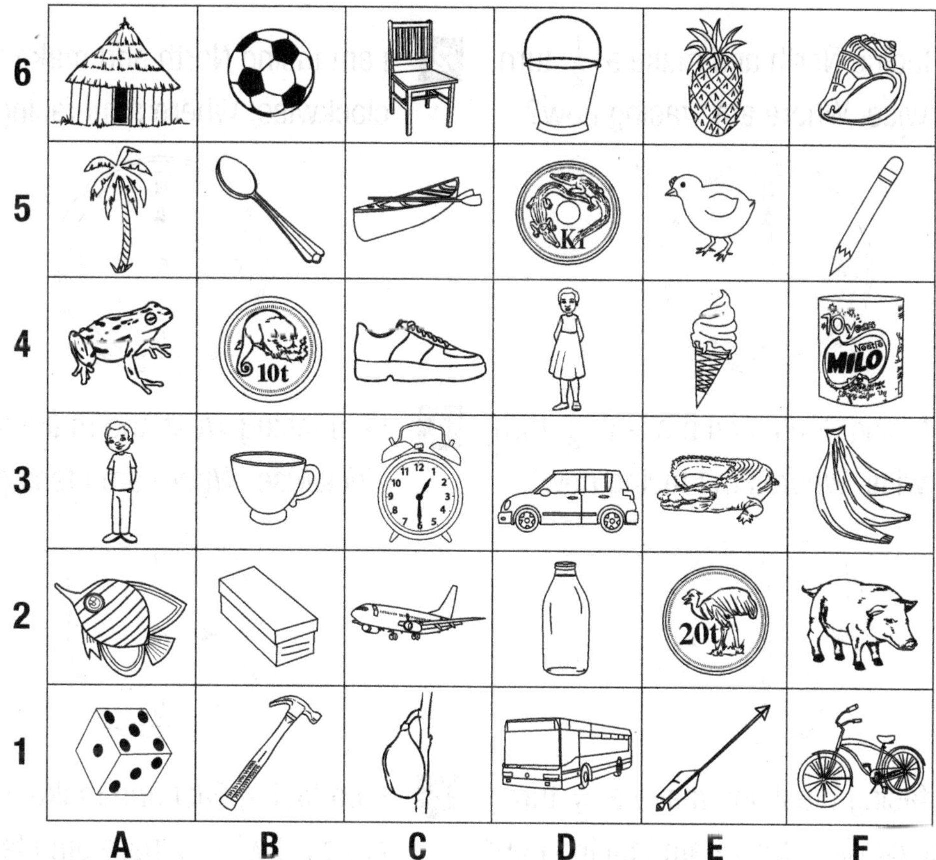

1 Use coordinates to describe the position of these objects.

a shell **b** shoe **c** palm tree **d** 20t coin

e ice cream cone **f** hammer **g** chair **h** crocodile

i canoe **j** bananas **k** K1 coin **l** bottle

m fish **n** bicycle **o** girl **p** chicken

q plane **r** pineapple

2 Name the objects at the following positions.

a B4 **b** F2 **c** B6 **d** C3 **e** E1 **f** A3

g D6 **h** F5 **i** A1 **j** F4 **k** A4 **l** B5

Use compass directions

1 I am facing North and make a $\frac{1}{2}$ turn clockwise. Where am I facing now?

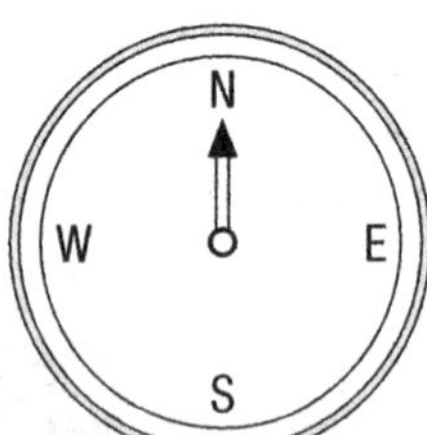

2 I am facing North and make a $\frac{1}{4}$ turn clockwise. Where am I facing now?

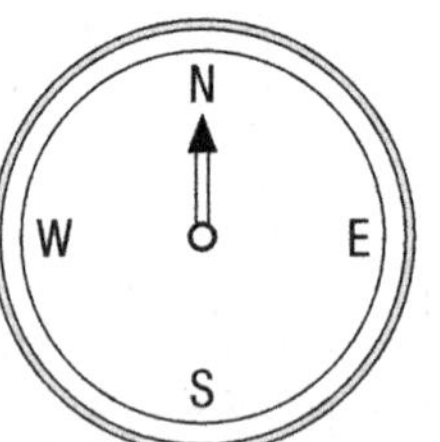

3 I am facing South and make a $\frac{1}{4}$ turn clockwise. Where am I facing now?

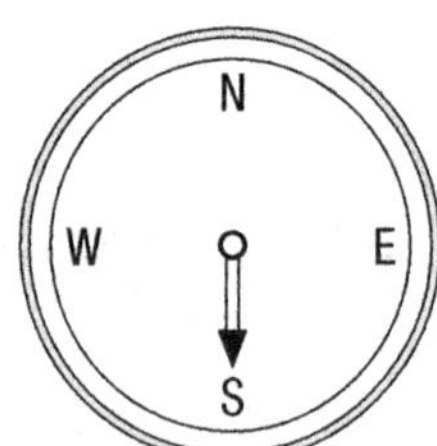

4 I am facing West and make a $\frac{3}{4}$ turn clockwise. Where am I facing now?

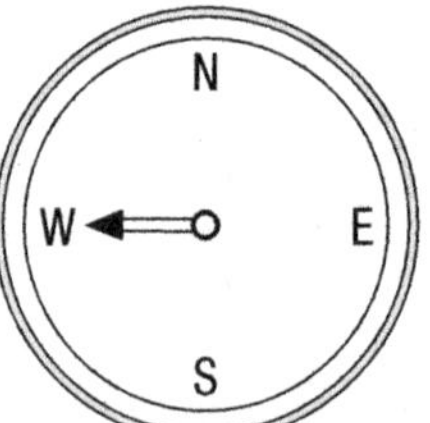

5 I am facing West and make a $\frac{1}{2}$ turn anticlockwise. Where am I facing now?

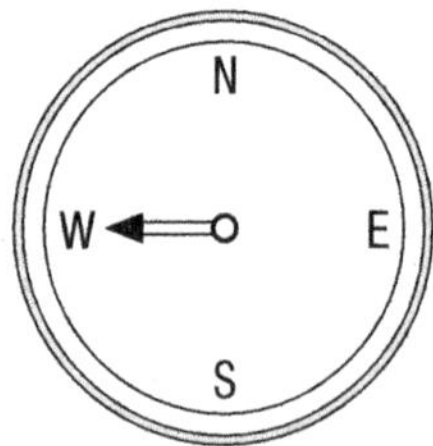

6 I am facing East and make a $\frac{1}{4}$ turn anticlockwise. Where am I facing now?

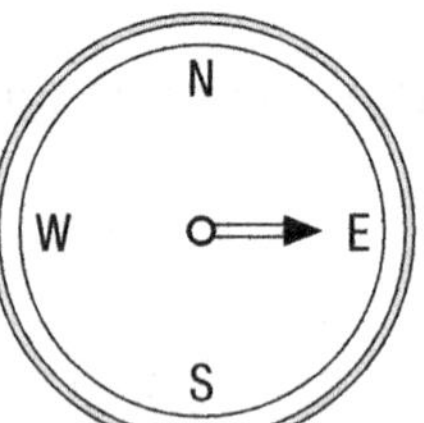

7 I am facing South and make a $\frac{3}{4}$ turn anticlockwise. Where am I facing now?

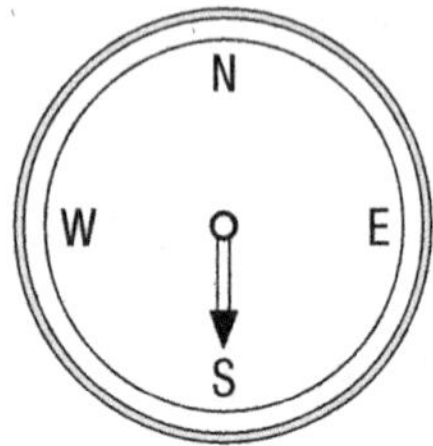

8 I am facing East and make a full turn clockwise. Where am I facing now?

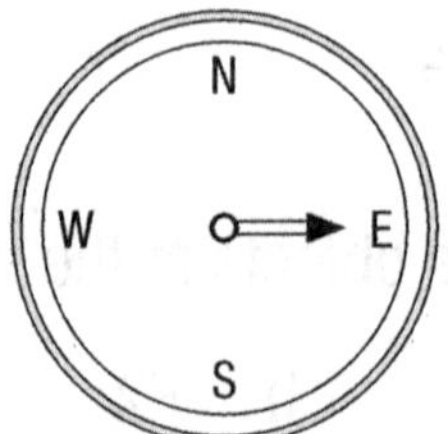

Assessment — Space and Shape

1 Name each of these shapes.

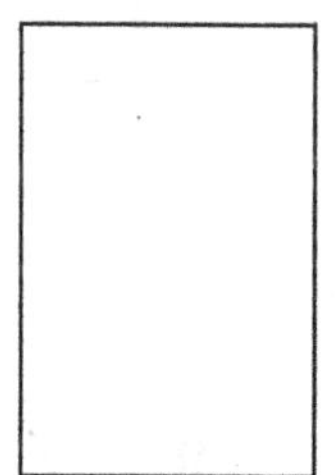

b

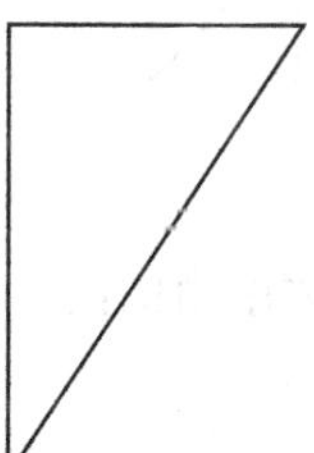

c

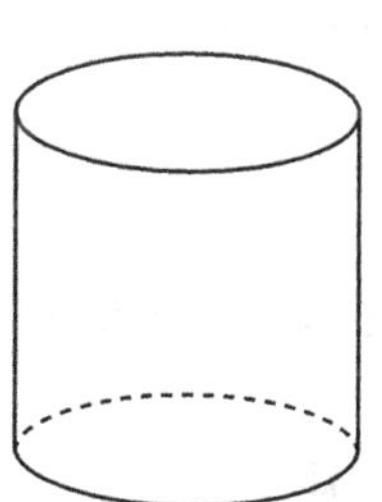

d

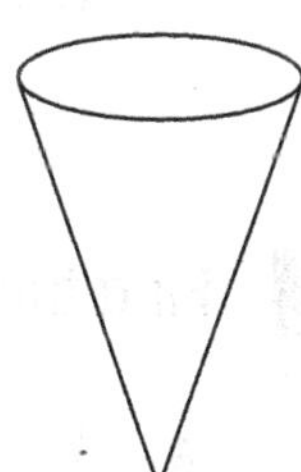

2 Draw another shape that would fit into each group.

a

b

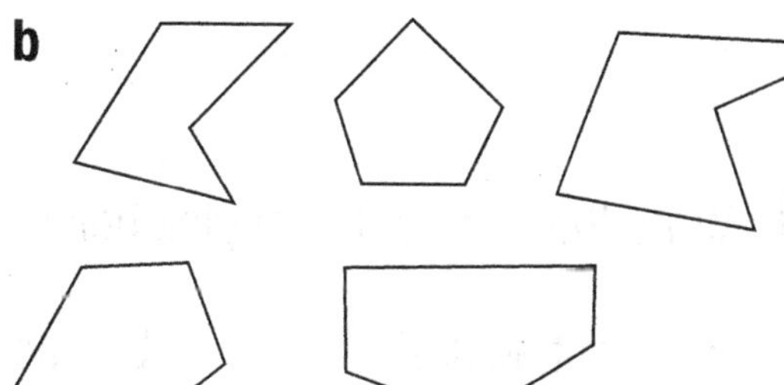

3 Write the letter of the shape below that has:

a 6 vertices **b** 6 faces **c** 8 edges

A

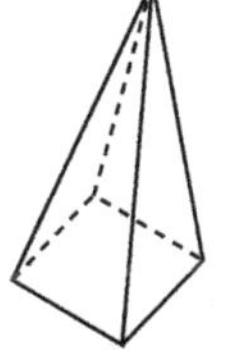

B

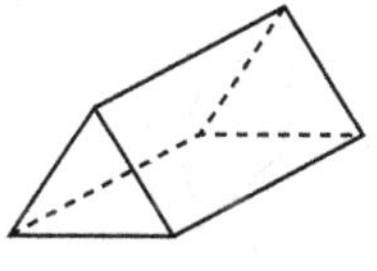

C

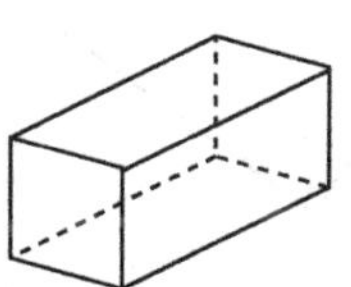

D

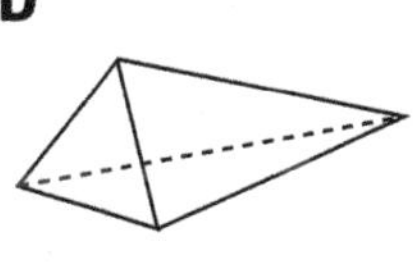

E

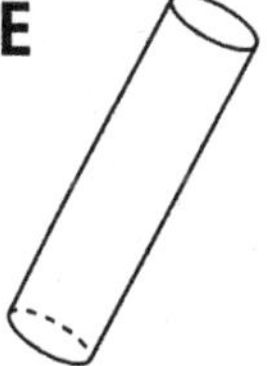

4 Copy these shapes. How many lines of symmetry does each shape have?

a

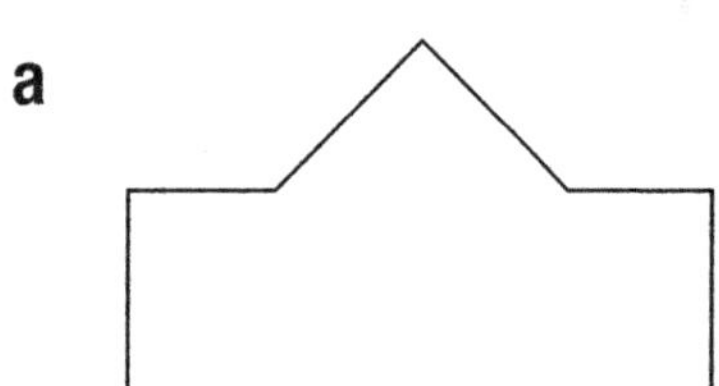

b

c

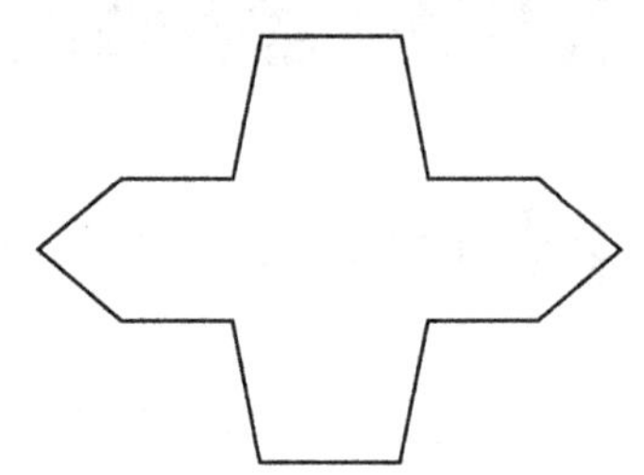

5 Find the angles that are more than a 'quarter turn'.

a

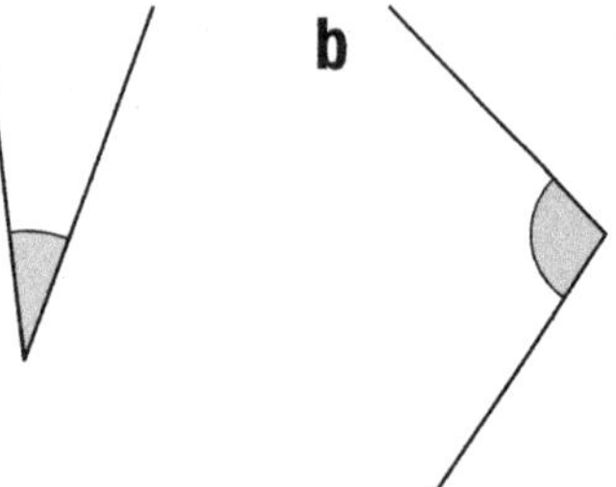

b

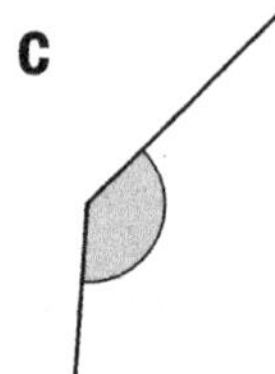

c

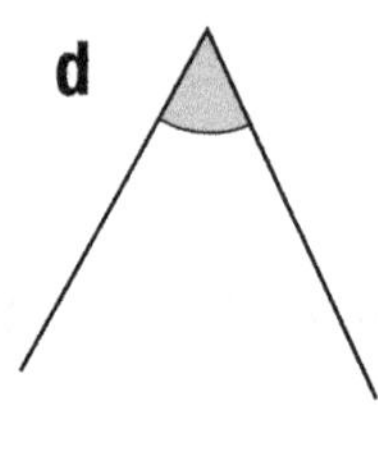

d

e

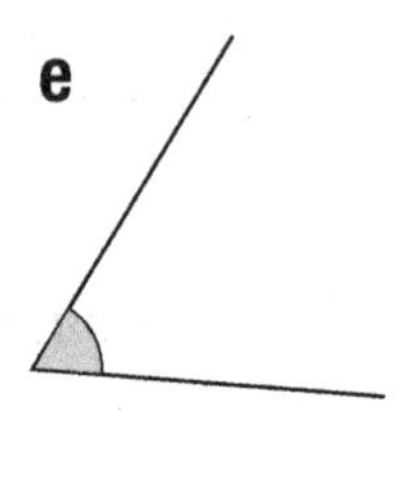

6 Write the letter of the angles below that are:

a right angles **b** acute angles **c** obtuse angles

A

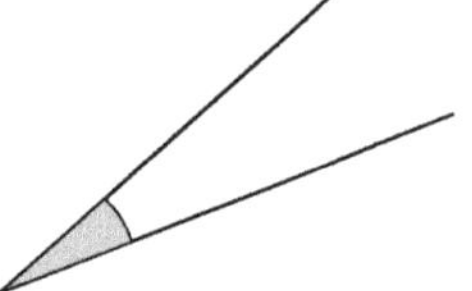

B

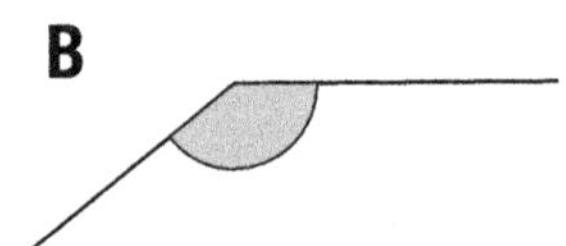

C

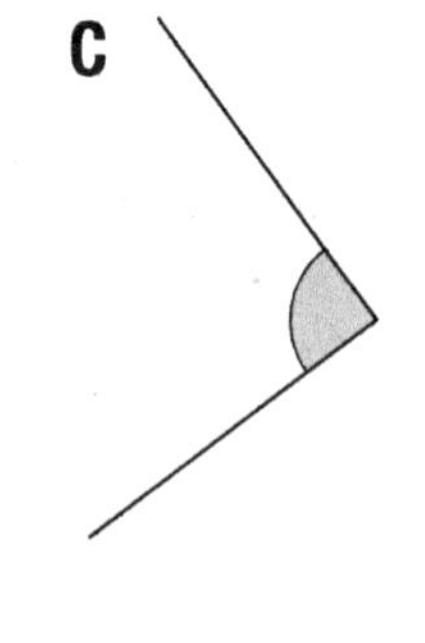

D

E

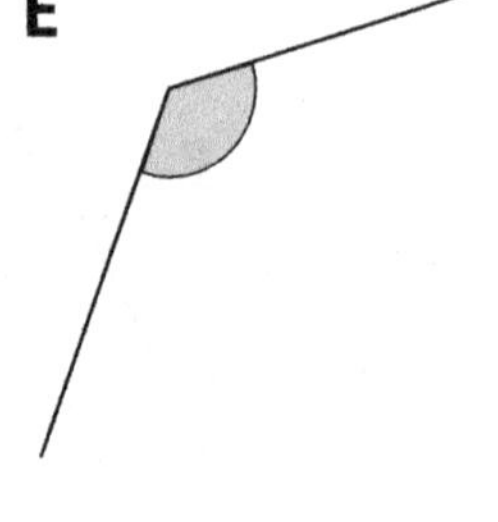

F

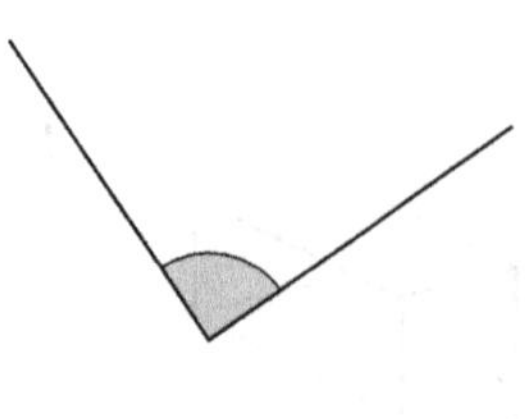

G

H

7 Copy this grid and follow the instructions.

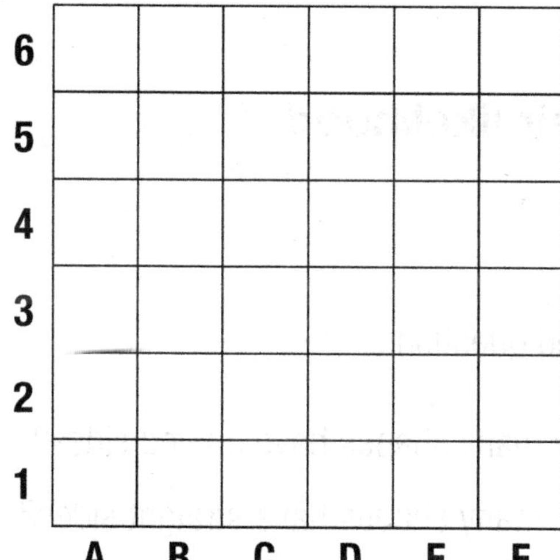

a Draw a cross (×) in D3.

b Draw a circle in B5.

c Draw a triangle in E6.

d Draw a rectangle in B2.

e Draw a square in A4.

f Write the first letter of your name in F1.

8 a I am facing East and make a $\frac{1}{2}$ turn clockwise. Where am I facing now?

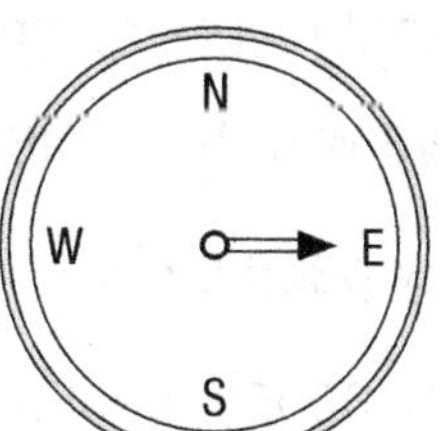

b I am facing West and make a $\frac{3}{4}$ turn anticlockwise. Where am I facing now?

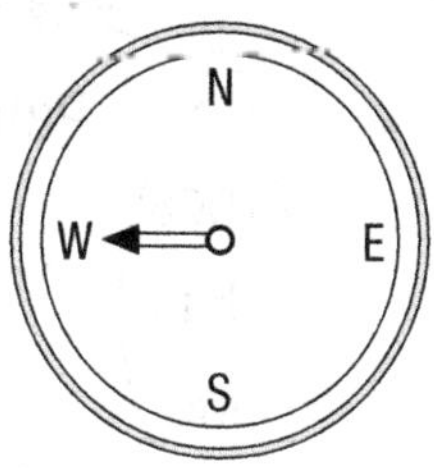

Strand Chance and Data

Compare events according to their likelihood

Read Venn diagrams

Look at each Venn diagram and answer the related questions.

1

a How many shapes have curved sides?

b How many shapes have straight sides?

c How many shapes have both curved and straight sides?

d Copy the diagram and draw another shape in each set.

2

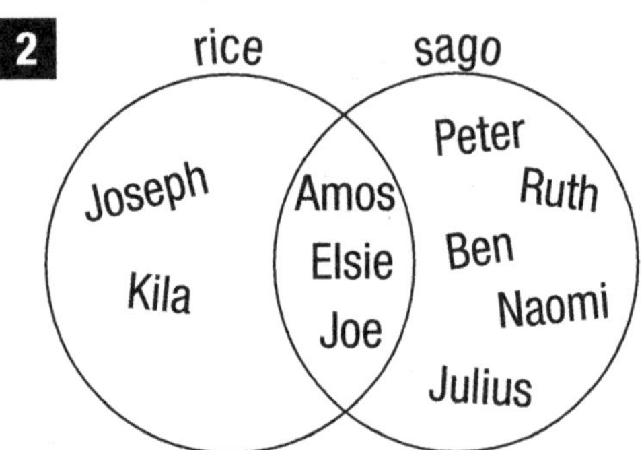

At a lunch, people were given the choice of rice or sago to have with their meat.

a How many people chose rice?

b How many people chose both rice and sago?

c How many people were at the lunch?

3

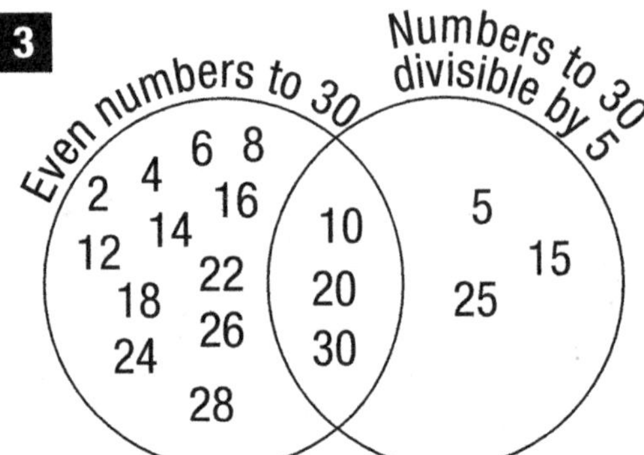

a Which numbers are both even and divisible by 5?

b How many numbers up to and including 30 are divisible by 5?

c How many numbers to 30 are not divisible by 5?

Construct and interpret information using graphs and simple timetables

Read bar graphs

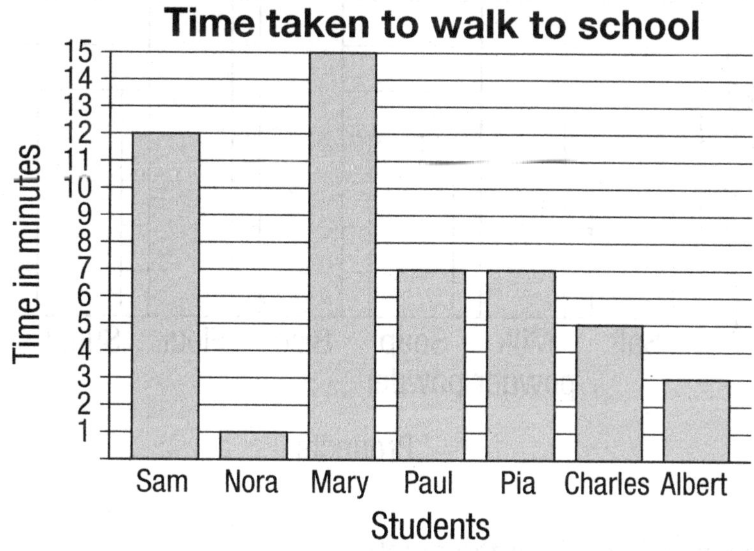

1 If all students walk at approximately the same speed:

a Who lives closest to school?

b Who lives furthest away?

c Which two students take the same time to walk to school?

d How much longer does it take Mary to walk to school than Charles?

e How long does it take Sam to walk to school?

f How much longer does it take Sam to walk to school than Paul?

g If Pia and Albert leave at the same time, who will arrive at school first?

h If Charles and Paul leave at the same time, who will arrive at school first?

i If Mary leaves for school at 8:15, what time will she arrive?

j If Charles leaves for school at 8 o'clock, what time will he arrive?

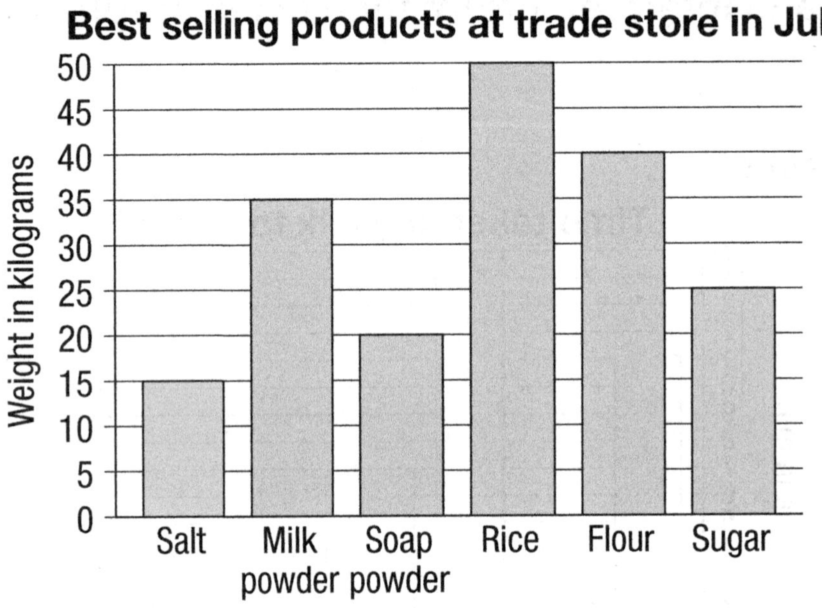

2 Answer these questions about the graph.

a Which product did the trade store sell the most of in July?

b How many kilograms of flour were sold?

c How much more milk powder than salt was sold?

d How much more rice than flour was sold?

e If all these products were loaded onto one truck, what would be their total weight?

f How many more kilograms of soap powder needed to be sold to match the best-selling product?

g If the trade store owner makes 20t on every kilogram of rice that is sold, how much did he make in July?

h If he makes 5t on every kilogram of salt that is sold, how much did he make?

Read a picture graph

Fish caught during one week

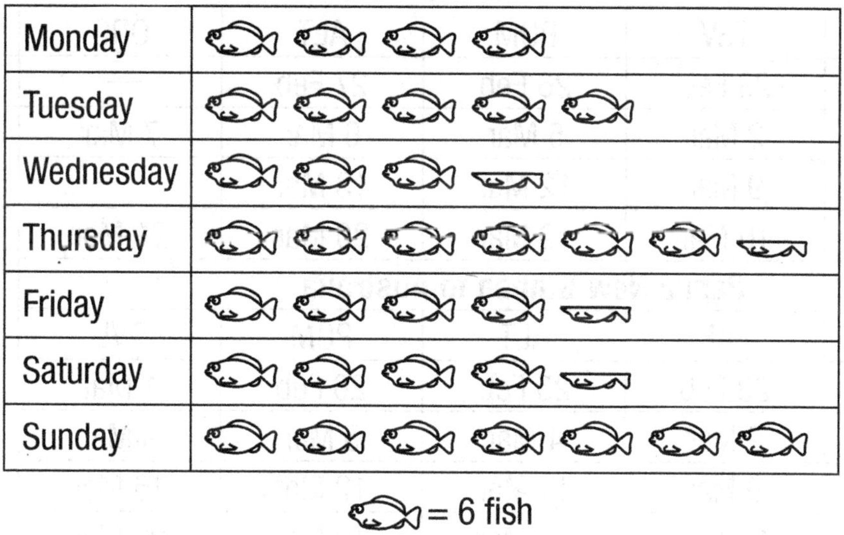

Monday	
Tuesday	
Wednesday	
Thursday	
Friday	
Saturday	
Sunday	

= 6 fish

1 How many fish were caught on Tuesday?

2 How many fish were caught on Friday?

3 Which day were the most fish caught?

4 What was the total number of fish caught on Saturday and Sunday?

5 How many more fish were caught on Thursday than on Wednesday?

6 How many more fish were caught on Sunday than on Friday?

7 On which days were an odd number of fish caught?

8 On which two days were the same number of fish caught?

9 How many fewer fish were caught on Wednesday than on Sunday?

10 What was the total number of fish caught on Monday, Tuesday and Wednesday?

11 If  = 4, how many fish were caught on these days?

a Tuesday **b** Sunday **c** Thursday

Read a timetable

Australia to Papua New Guinea					
Vessel	TSV	POM	ALT	ORO	LAE
1	23 Feb	26 Feb	27 Feb	—	1 Mar
2	2 Mar	5 Mar	6 Mar	7 Mar	8 Mar
3	9 Mar	12 Mar	13 Mar	—	15 Mar
4	16 Mar	19 Mar	20 Mar	21 Mar	22 Mar
Papua New Guinea to Australia					
Vessel	LAE	ALT	POM	TVL	
2	23 Feb	25 Feb	26 Feb	1 Mar	
3	2 Mar	4 Mar	5 Mar	9 Mar	
1	9 Mar	11 Mar	12 Mar	16 Mar	
2	16 Mar	18 Mar	19 Mar	23 Mar	

(TSV = Townsville, POM = Port Moresby, ALT = Alotau, ORO = Oro Province)

1 What date does Vessel 1 depart Townsville in Australia for its trip to Papua New Guinea?

2 What date does Vessel 1 arrive in Lae on its trip from Australia?

3 How many days does Vessel 1 have in Lae before beginning its return journey to Australia?

4 If all vessels depart in the morning and arrive at night, how many days is the trip from Lae back to Townsville?

5 According to the schedule on the previous page, how many trips does Vessel 2 make between Australia and Papua New Guinea?

6 From which country does Vessel 2 begin its first trip?

7 What date does Vessel 4 arrive in Alotau?

8 If Vessel 3 departs Lae in the morning and arrives in Port Moresby at night, how many days is its journey between the two places?

9 Does the trip from Townsville to Port Moresby or the trip from Port Moresby to Townsville take longer? Why do you think this is?

10 If a vessel left Lae on the 7th May, what date should it arrive in Townsville?

Assessment — Chance and Data

1 Mr Siwi graphed the results from a survey about the most popular colours.

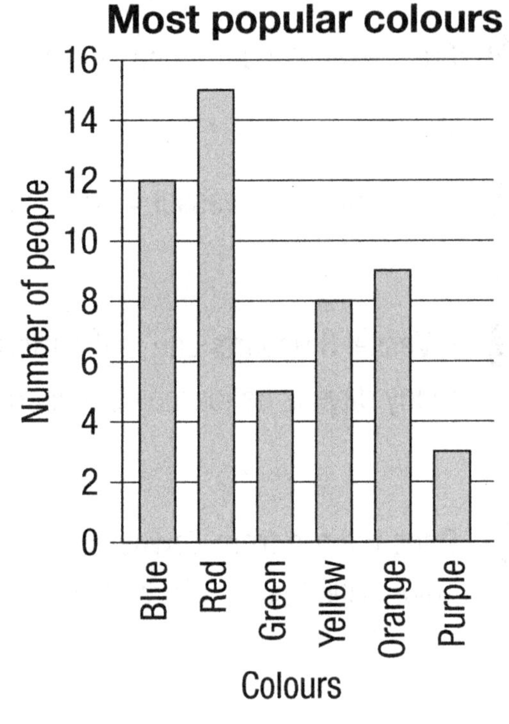

- **a** What was the most popular colour?
- **b** How many people chose blue?
- **c** How many people chose green?
- **d** How many more people chose red than blue?
- **e** How many more people chose blue than green?
- **f** How many fewer people chose orange than red?
- **g** How many people were surveyed?

2

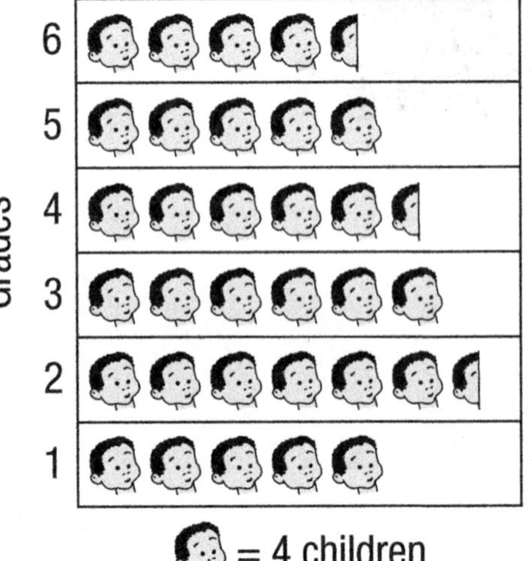

- **a** How many children are in grade 3?
- **b** How many children are in grade 1?
- **c** How many children are in grade 6?
- **d** Which is the smallest class?
- **e** Which is the largest class?
- **f** Which grade has 22 children?
- **g** How many more children are in grade 2 than in grade 4?
- **h** What is the total number of children in grades 3 and 4?
- **i** How many children are at the school?

Strand Patterns

4.5.1 Develop arithmetical rules to describe number patterns

Recognise and complete number patterns

1 Copy and complete the number charts.

a

3512	3514	3516	3518	3520
3522		3526	3528	
				3540
			3548	

b

5125	5225	5325
5425		
	5825	
6025		

c

1841	1851	1861		1881		1901	1911	1921	1931
1941	1951		1971		1991	2001	2011		
		2061	2071	2081		2101			
2141					2191		2211		2231
	2251		2271					2321	
2341				2381		2401			

d

2067	2072	2077	2082
2087		2097	2102
	2112	2117	
2127			2142
2147			
	2172		2182
		2197	2202
2207			
	2232		2242

e

4763	4863	4963	5063
5163			
	5663		5863
	6063	6163	
6363	6463		
		6963	7063
7163			
	7663		
7963		8163	

2 Copy and complete these number sequences by adding or subtracting the number at the start.

a	**+ 100**	1682	______	______	______	______	______
b	**– 10**	4561	______	______	______	______	______
c	**+ 5**	2091	______	______	______	______	______
d	**+ 10**	7477	______	______	______	______	______
e	**– 100**	3214	______	______	______	______	______
f	**+ 2**	1996	______	______	______	______	______
g	**– 5**	8113	______	______	______	______	______
h	**– 10**	5623	______	______	______	______	______
i	**+ 100**	2758	______	______	______	______	______
j	**+ 5**	3816	______	______	______	______	______
k	**– 2**	9553	______	______	______	______	______
l	**+ 1**	4998	______	______	______	______	______
m	**+ 10**	1985	______	______	______	______	______
n	**– 20**	5000	______	______	______	______	______
o	**+ 50**	6890	______	______	______	______	______
p	**– 1**	7002	______	______	______	______	______
q	**+ 20**	8459	______	______	______	______	______

r – 200 2460 ______ ______ ______ ______ ______

s – 50 3210 ______ ______ ______ ______ ______

t + 500 1600 ______ ______ ______ ______ ______

u – 100 5422 ______ ______ ______ ______ ______

v – 5 4074 ______ ______ ______ ______ ______

w + 20 6975 ______ ______ ______ ______ ______

x – 500 9650 ______ ______ ______ ______ ______

Identify rules and complete number sequences

Copy and complete the number sequences below and record the rule for each one. (For example, the rule for the sequence 564, 562, 560, 558, 556 is 'subtract 2'.)

1 731, 721, 711, 701, 691, ______, ______, ______, ______, ______

2 540, 580, 620, 660, 700, ______, ______, ______, ______, ______

3 811, 816, 821, 826, 831, ______, ______, ______, ______, ______

4 608, 605, 602, 599, 596, ______, ______, ______, ______, ______

5 1200, 1220, 1240, 1260, 1280, ______, ______, ______, ______, ______

6 3446, 3441, 3436, 3431, 3426, ______, ______, ______, ______, ______

7 7009, 7007, 7005, 7003, 7001, ______, ______, ______, ______, ______

8. 5675, 5685, 5695, 5705, 5715, ______, ______, ______, ______, ______

9. 2657, 2664, 2671, 2678, 2685, ______, ______, ______, ______, ______

10. 1055, 1035, 1015, 995, 975, ______, ______, ______, ______, ______

11. 7710, 7810, 7910, 8010, 8110, ______, ______, ______, ______, ______

12. 4360, 4310, 4260, 4210, 4160, ______, ______, ______, ______, ______

13. 11.5, 12, 12.5, 13, 13.5, ______, ______, ______, ______, ______

14. 12, 14.5, 17, 19.5, 22, ______, ______, ______, ______, ______

15. 25, 23.5, 22, 20.5, 19, ______, ______, ______, ______, ______

16. 15, 13.75, 12.5, 11.25, 10, ______, ______, ______, ______, ______

17. 36, 36.25, 36.5, 36.75, 37, ______, ______, ______, ______, ______

18. 16, $15\frac{1}{2}$, 15, $14\frac{1}{2}$, 14, ______, ______, ______, ______, ______

19. 5, $5\frac{1}{4}$, $5\frac{1}{2}$, $5\frac{3}{4}$, 6, ______, ______, ______, ______, ______

20. 10, $9\frac{1}{4}$, $8\frac{1}{2}$, $7\frac{3}{4}$, 7, ______, ______, ______, ______, ______

21. 20, $18\frac{1}{2}$, 17, $15\frac{1}{2}$, 14, ______, ______, ______, ______, ______

22. 32, $34\frac{1}{4}$, $36\frac{1}{2}$, $38\frac{3}{4}$, 41, ______, ______, ______, ______, ______

Match number patterns to rules

Match each pattern in the left column to its rule in the right column by recording the number of the pattern and the letter of the matching rule.

Pattern	Rule
1 164, 168, 172, 176, 180 …	**a** Add 2.5
2 817, 811, 805, 799, 793 …	**b** Subtract $\frac{1}{2}$
3 623, 611, 599, 587, 575 …	**c** Subtract 7.5
4 1167, 1174, 1181, 1188, 1195 …	**d** Subtract $1\frac{1}{2}$
5 5608, 5603, 5598, 5593, 5588 …	**e** Add 25
6 2510, 2525, 2540, 2555, 2570 …	**f** Subtract 6
7 37, 39.5, 42, 44.5, 47 …	**g** Add 4
8 45, $43\frac{1}{2}$, 42, $40\frac{1}{2}$, 39 …	**h** Add $1\frac{1}{4}$
9 85, 77.5, 70, 62.5, 55 …	**i** Add 15
10 22, $23\frac{1}{4}$, $24\frac{1}{2}$, $25\frac{3}{4}$, 27 …	**j** Subtract 12
11 74, $73\frac{1}{2}$, 73, $72\frac{1}{2}$, 72 …	**k** Subtract 5
12 3650, 3675, 3700, 3725, 3750 …	**l** Add 7

Create and identify number patterns

1 Write a sequence that has at least 8 numbers for each rule. Start at:

a 1500 and keep adding 7

b 850 and keep subtracting 8

c 35 and keep subtracting $\frac{1}{2}$

d 2345 and keep adding 10

e 7233 and keep subtracting 5

f 4305 and keep subtracting 100

g 22 and keep adding 2.5

h 17 and keep adding 0.5

i 2 and keep multiplying by 2

j 6400 and keep dividing by 2

k 3124 and keep subtracting 10

l 5870 and keep adding 20

m 1598 and keep adding 100

n 2446 and keep subtracting 4

o 50 and keep subtracting $2\frac{1}{2}$

p 40 and keep adding $1\frac{1}{4}$

q 36 and keep adding 1.5

r 15 and keep subtracting 0.25

s 6755 and keep subtracting 50

t 3800 and keep adding 25

u 1.5 and keep multiplying by 2

v $\frac{1}{4}$ and keep multiplying by 2

w 70 and keep subtracting 4.5

x 28 and keep adding $\frac{3}{4}$

y 4976 and keep adding 5

z 7150 and keep subtracting 25

2 Record the rule that has been used to create each number pattern.

a 2500, 2450, 2400, 2350, 2300, 2250, 2200, 2150

b 24, 21.5, 19, 16.5, 14, 11.5, 9, 6.5, 4, 1.5

c 12, 19.5, 27, 34.5, 42, 49.5, 57, 64.5, 72, 79.5

d $\frac{3}{4}$, $1\frac{1}{2}$, 3, 6, 12, 24, 48, 96, 192, 384, 768

e 50, $51\frac{1}{2}$, 53, $54\frac{1}{2}$, 56, $57\frac{1}{2}$, 59, $60\frac{1}{2}$, 62, $63\frac{1}{2}$, 65

f 9600, 4800, 2400, 1200, 600, 300, 150, 75, 37.5

g 3411, 3404, 3397, 3390, 3383, 3376, 3369, 3362

h 1086, 1101, 1116, 1131, 1146, 1161, 1176, 1191

i 5650, 5630, 5610, 5590, 5570, 5550, 5530, 5510

j 700, 900, 1100, 1300, 1500, 1700, 1900, 2100

k 41, 42.5, 44, 45.5, 47, 48.5, 50, 51.5, 53, 54.5

l 65, 64.5, 64, 63.5, 63, 62.5, 62, 61.5, 61, 60.5

m 36, $35\frac{3}{4}$, $35\frac{1}{2}$, $35\frac{1}{4}$, 35, $34\frac{3}{4}$, $34\frac{1}{2}$, $34\frac{1}{4}$, 34

n 2380, 2880, 3380, 3880, 4380, 4880, 5380, 5880

o 8300, 8150, 8000, 7850, 7700, 7550, 7400, 7250

p 4700, 4950, 5200, 5450, 5700, 5950, 6200, 6450

Assessment Patterns

1 Copy and complete these number charts.

a

5881	5886	5891	5896
5901		5911	
	5926		

b

3736	3836	3936	4036
	4236		
4536			

2 a Complete this number sequence by subtracting 20 each time.

3154 ______ ______ ______ ______ ______ ______ ______ ______

b Complete this number sequence by adding 50 each time.

1869 ______ ______ ______ ______ ______ ______ ______ ______

c Complete this number sequence by subtracting 1.5 each time.

75 ______ ______ ______ ______ ______ ______ ______ ______

3 Copy and complete these number sequences and record the rule for each one.

a 476, 480, 484, 488, 492, ______, ______, ______, ______, ______, ______

b 613, 603, 593, 583, 573, ______, ______, ______, ______, ______, ______

c 8144, 8094, 8044, 7994, 7944, ______, ______, ______, ______, ______

d 5787, 5807, 5827, 5847, 5867, ______, ______, ______, ______, ______

4 Record the rule that has been used to create each number pattern.

a 36, 39.5, 43, 46.5, 50, 53.5, 57, 60.5, 64, 67.5

b 45, $43\frac{1}{2}$, 42, $40\frac{1}{2}$, 39, $37\frac{1}{2}$, 36, $34\frac{1}{2}$, 33, $31\frac{1}{2}$

c 50, $50\frac{3}{4}$, $51\frac{1}{2}$, $52\frac{1}{4}$, 53, $53\frac{3}{4}$, $54\frac{1}{2}$, $55\frac{1}{4}$, 56, $56\frac{3}{4}$

5 For each of the following rules, write a sequence that has at least 8 numbers.

a Start at 1250 and keep subtracting 25.

b Start at 60 and keep adding 2.5.

c Start at 4816 and keep subtracting 7.

Important Facts

Number

Times tables

These are the times tables you should know by the end of grade 4.

1 × table	2 × table	3 × table	4 × table
1 × 0 = 0	2 × 0 = 0	3 × 0 = 0	4 × 0 = 0
1 × 1 = 1	2 × 1 = 2	3 × 1 = 3	4 × 1 = 4
1 × 2 = 2	2 × 2 = 4	3 × 2 = 6	4 × 2 = 8
1 × 3 = 3	2 × 3 = 6	3 × 3 = 9	4 × 3 = 12
1 × 4 = 4	2 × 4 = 8	3 × 4 = 12	4 × 4 = 16
1 × 5 = 5	2 × 5 = 10	3 × 5 = 15	4 × 5 = 20
1 × 6 = 6	2 × 6 = 12	3 × 6 = 18	4 × 6 = 24
1 × 7 = 7	2 × 7 = 14	3 × 7 = 21	4 × 7 = 28
1 × 8 = 8	2 × 8 = 16	3 × 8 = 24	4 × 8 = 32
1 × 9 = 9	2 × 9 = 18	3 × 9 = 27	4 × 9 = 36
1 × 10 = 10	2 × 10 = 20	3 × 10 = 30	4 × 10 = 40
1 × 11 = 11	2 × 11 = 22	3 × 11 = 33	4 × 11 = 44
1 × 12 = 12	2 × 12 = 24	3 × 12 = 36	4 × 12 = 48

5 × table	6 × table	10 × table	11 × table
5 × 0 = 0	6 × 0 = 0	10 × 0 = 0	11 × 0 = 0
5 × 1 = 5	6 × 1 = 6	10 × 1 = 10	11 × 1 = 11
5 × 2 = 10	6 × 2 = 12	10 × 2 = 20	11 × 2 = 22
5 × 3 = 15	6 × 3 = 18	10 × 3 = 30	11 × 3 = 33
5 × 4 = 20	6 × 4 = 24	10 × 4 = 40	11 × 4 = 44
5 × 5 = 25	6 × 5 = 30	10 × 5 = 50	11 × 5 = 55
5 × 6 = 30	6 × 6 = 36	10 × 6 = 60	11 × 6 = 66
5 × 7 = 35	6 × 7 = 42	10 × 7 = 70	11 × 7 = 77
5 × 8 = 40	6 × 8 = 48	10 × 8 = 80	11 × 8 = 88
5 × 9 = 45	6 × 9 = 54	10 × 9 = 90	11 × 9 = 99
5 × 10 = 50	6 × 10 = 60	10 × 10 = 100	11 × 10 = 110
5 × 11 = 55	6 × 11 = 66	10 × 11 = 110	11 × 11 = 121
5 × 12 = 60	6 × 12 = 72	10 × 12 = 120	11 × 12 = 132

Place value

The value of a digit in a number is determined by its place value or position. The decimal point is used to separate the whole number part from the fraction part.

Thousands	Hundreds	Tens	Ones	Tenths	Hundredths

Measurement

Length and perimeter

Millimetres, centimetres and metres are used to measure length.

10 millimetres (mm) = 1 centimetre (cm) $1 \text{ mm} = \frac{1}{10} \text{ cm} = 0.1 \text{ cm}$

100 centimetres (cm) = 1 metre (m) $1 \text{ cm} = \frac{1}{100} \text{ m} = 0.01 \text{ m}$

Capacity

Millilitres and litres are used to measure capacity.

1000 millilitres (mL) = 1 litre (L)

Weight

Grams and kilograms are used to measure weight.

1000 grams (g) = 1 kilogram (kg)

Money

100 toea (t) = 1 kina (K) $1\text{t} = \frac{1}{100} \text{ K} = \text{K}0.01$

Volume

Cubic centimetres and cubic metres are used to measure volume.

100 000 cubic centimetres (cm^3) = 1 cubic metre (m^3)

Time

60 seconds = 1 minute

60 minutes = 1 hour

24 hours = 1 day

7 days = 1 week

2 weeks = 1 fortnight

52 weeks = 1 year

12 months = 1 year

365 days = 1 year

366 days = 1 leap year

Answers

Strand: Number and Application

Pages 2–5

1
- a K809 = eight hundred and nine kina
- b K574 = five hundred and seventy-four kina
- c K290 = two hundred and ninety kina
- d K156 = one hundred and fifty-six kina
- e K2667 = two thousand, six hundred and sixty-seven kina
- f K5012 = five thousand and twelve kina
- g K3805 = three thousand, eight hundred and five kina
- h K6230 = six thousand, two hundred and thirty kina

2
- a four hundred and sixteen
- b eight hundred and forty-five
- c one hundred and seventy
- d six hundred and thirty-eight
- e nine hundred and eighty-two
- f three thousand, eight hundred and five
- g four thousand and thirty-three
- h seven thousand, four hundred and eighteen
- i one thousand, nine hundred and sixty-one
- j six thousand, two hundred and seventy-four
- k five thousand, four hundred and ten
- l nine thousand, eight hundred and fifty-five
- m two thousand and six
- n eight thousand, seven hundred and twenty-four
- o one thousand, two hundred and seventy-nine

3 a 517 b 235 c 841 d 409 e 128 f 763
g 6474 h 9708 i 3085 j 8519 k 2657 l 7642

4 a 875 b 360 c 123 d 633 e 401 f 517
g 1047 h 9025 i 5266 j 2559 k 9322 l 7505

5
- a six hundred and twenty-three
- b eight hundred and eighty-two
- c three hundred and nine
- d four hundred and fifty-five
- e nine hundred and sixteen
- f six thousand, eight hundred and eighty-two
- g two thousand, two hundred and seventy-one
- h four thousand, three hundred and seventy-seven
- i nine thousand, six hundred and fifty
- j one thousand, seven hundred and nine
- k three thousand, two hundred and twenty-eight
- l eight thousand, eight hundred and fifty-nine
- m five thousand, one hundred and eighty
- n seven thousand, nine hundred and thirty-six
- o nine thousand, three hundred and sixty-four

6 a 216 b 570 c 784 d 901 e 421 f 159
g 6048 h 5009 i 8735 j 3120 k 9562 l 1378

Pages 5–6

1 a – l Answers will vary.

2 a – l Answers will vary.

3 a < b > c < d < e > f >
g > h < i < j < k < l >
m > n > o < p > q > r <

4 a – u Answers will vary.

Pages 7–9

1 a 757 b 684 c 919 d 530 e 747 f 310
g 561 h 983 i 662 j 551 k 877 l 746
m 183 n 392 o 771 p 009 q 354 r 072
s 565 t 870 u 253 v 155 w 481 x 765
y 963 z 383

2 a 4225 b 6179 c 2358 d 1040 e 7269 f 5833
g 5409 h 3618 i 1259 j 8014 k 2060 l 4779
m 9256 n 6044 o 3415 p 5004 q 7898 r 1228
s 4455 t 9050 u 5216 v 8568 w 2227 x 3222

3 a – o Answers will vary.

4 a – o Answers will vary.

5
- a K410, K800, K3500, K3750, K4000, K4150
- b K650, K825, K965, K1280, K1670, K1855
- c K260, K1030, K1130, K2030, K2160, K2300
- d K525, K590, K1550, K5090, K5250, K5500
- e K870, K895, K8070, K8075, K8700, K8950
- f K1350, K1730, K3085, K3100, K3175, K3350
- g K707, K775, K5700, K7050, K7110, K7500
- h K890, K908, K8090, K8990, K9080, K9880
- i K460, K466, K4060, K4160, K4600, K4660
- j K990, K995, K1019, K1090, K1190, K1990

Page 10

1 a 304 b 540 c 85 d 1562 e 2178 f 1406

2 a 2251 b 4713 c 3253 d 1037 e 8204 f 4460

Page 11

1 548 **2** 369 **3** 675 **4** 122 **5** 816
6 237 **7** 450 **8** 391 **9** 183 **10** 705
11 2553 **12** 5827 **13** 1326 **14** 2615 **15** 7804
16 4380 **17** 6951 **18** 1582

Page 12

1 a 521 and 125 b 865 and 568 c 743 and 347
d 320 and 203 e 966 and 669 f 982 and 289

2 a 7631 and 1367 b 8752 and 2578 c 9420 and 2049
d 8441 and 1448 e 6530 and 3056 f 9752 and 2579

3 a 942, 924, 492, 429, 249, 294 b 866, 668, 686

4
- a 4923, 4932, 4293, 4239, 4329, 4392, 9432, 9423, 9342, 9324, 9234, 9243, 3294, 3249, 3429, 3492, 3942, 3924, 2349, 2394, 2439, 2493, 2934, 2943
- b 9432, 9423, 9342, 9324, 9234, 9243
- c 4932, 4239, 9432, 9234, 2439, 2934

Page 13

1 216, 5267, 869, 3868

2 597, 5400, 560

3 635, 3776, 8312, 831

4 540, 341

5 755, 1738, 5764, 172, 2076

6 297, 1926, 3298, 698

7 652, 4261, 5233

8 6128, 153, 4614, 3018, 188

Pages 14–15

1 507, 514, 521, 528, 535, 542, 549, 556, 563, 570, 577, 584, 591

2 **a** 452, 460, 468, 476, 484, 492 **b** 203, 211, 219, 227, 235, 243
c 597, 605, 613, 621, 629, 637 **d** 780, 788, 796, 804, 812, 820
e 391, 399, 407, 415, 423, 431 **f** 618, 626, 634, 642, 650, 658
g 926, 934, 942, 950, 958, 966

3 **a** 195, 201, 207, 213, 219, 225 **b** 898, 904, 910, 916, 922, 928
c 623, 629, 635, 641, 647, 653 **d** 916, 922, 928, 934, 940, 946
e 741, 747, 753, 759, 765, 771 **f** 267, 273, 279, 285, 291, 297
g 509, 515, 521, 527, 533, 539

Page 16

1 **a** 2173, 2182, 2191, 2200, 2209, 2218
b 8979, 8988, 8997, 9006, 9015, 9024
c 1058, 1067, 1076, 1085, 1094, 1103
d 5114, 5123, 5132, 5141, 5150, 5159
e 9260, 9269, 9278, 9287, 9296, 9305
f 3727, 3736, 3745, 3754, 3763, 3772
g 7385, 7394, 7403, 7412, 7421, 7430

2 **a** 4238, 4245, 4252, 4259, 4266, 4273
b 7220, 7227, 7234, 7241, 7248, 7255
c 1655, 1662, 1669, 1676, 1683, 1690
d 6987, 6994, 7001, 7008, 7015, 7022
e 2979, 2986, 2993, 3000, 3007, 3014
f 5003, 5010, 5017, 5024, 5031, 5038
g 8874, 8881, 8888, 8895, 8902, 8909

Pages 17–18

1 **a** 30 **b** 38 **c** 42 **d** 27 **e** 44
f 39 **g** 30 **h** 21 **i** 36 **j** 42

2 **a**

+	32	27	49	18
9	41	36	58	27
7	39	34	56	25
12	44	39	61	30
20	52	47	69	38
30	62	57	79	48
25	57	52	74	43

b

+	39	55	74	46
8	47	63	82	54
6	45	61	80	52
11	50	66	91	57
15	54	70	89	61
19	58	74	93	65
29	68	84	103	75

3 **a** 100 **b** 50 **c** 65 **d** 95 **e** 95
f 65 **g** 53 **h** 95 **i** 79 **j** 76
k 93 **l** 100 **m** 106 **n** 88 **o** 67
p 88 **q** 62 **r** 86 **s** 70 **t** 45
u 79 **v** 85 **w** 87 **x** 74

4 Possible answers include:
a 50 + 15 + 35, 35 + 25 + 40 **b** 80 + 15 + 5, 65 + 25 + 10
c 50 + 30 + 20 **d** 37 + 23 + 40, 37 + 33 + 30
e 51 + 29 + 20 **f** 52 + 38 + 10, 38 + 22 + 40
g 44 + 26 + 30, 56 + 26 + 18 **h** 42 + 36 + 22

Pages 19–20

1 **a**
12 15
18 3 6 11
9 2
9
5 7
14 8 10 16
17 19

b
18 13
11 11 6 10
4 3
7
7 8
14 10 5 15
17 12

c
9 12
11 4 7 14
6 9
5
10 3
15 5 8 8
10 13

d
10 17
15 4 11 13
9 7
6
6 10
12 8 5 16
14 11

e
11 16
15 3 8 20
7 12
8
10 5
18 4 11 13
12 19

f
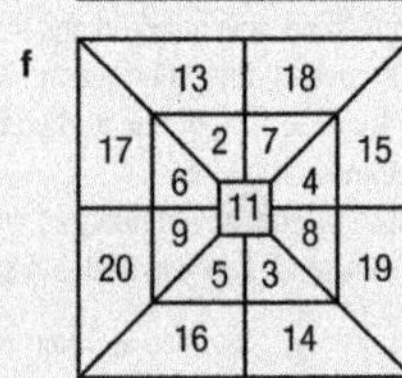

2 **a**
7
8 1
29 21 20
35 6 27 7

b
13
22 9
28 6 15
36 8 2 17

c
4
27 31
33 6 37
41 8 2 39

d
15
27 12
31 4 16
36 5 9 25

e
4
38 34
41 3 37
47 6 3 40

f
15
32 47
38 6 53
51 13 7 60

g
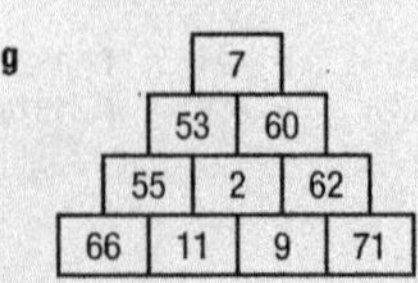

h
18
35 53
39 4 57
51 12 8 65

Page 21

1 165, 174, 168, 173, 184, 175

2 798, 808, 802, 793, 798, 813

3 503, 510, 499, 494, 504, 512, 521

4 2010. 1999, 2002, 1994, 1989, 1999

5 1490, 1502, 1497, 1491, 1495, 1503, 1500, 1505

6 5686, 5680, 5672, 5676, 5671, 5677, 5670

Page 22

Game

Pages 23–26

1 a

12, 36, 27, 15, 21, 9, 18, 24
4, 12, 9, 5, 7, 3, 6, 8
× 3

b

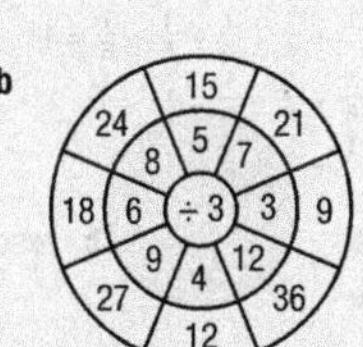

c

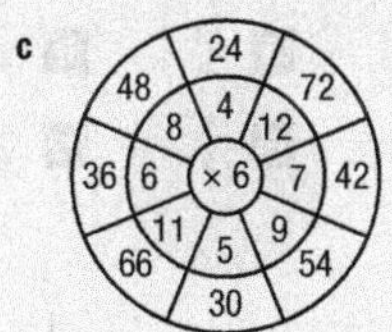

d

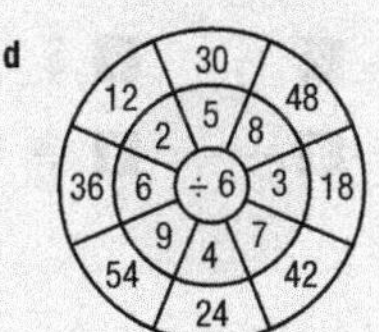

e

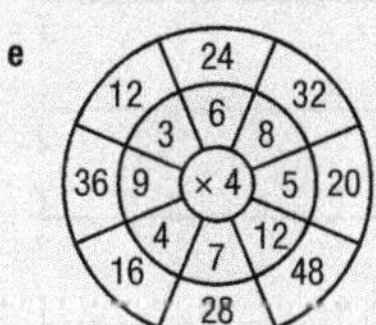

f

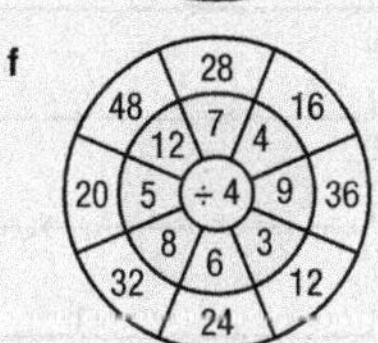

g

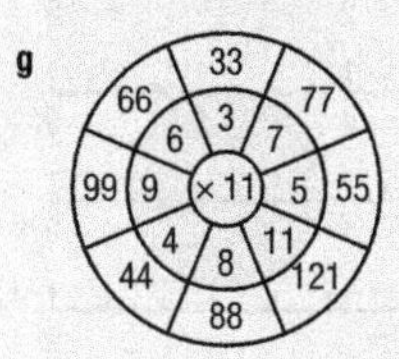

h

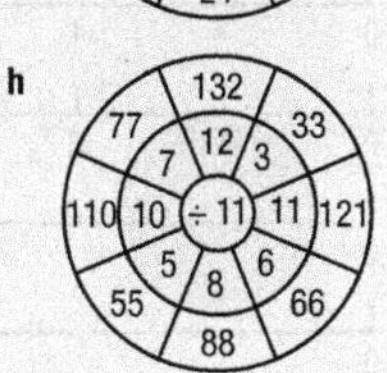

2 a 7, 3, 10, 5, 8, 4, 9, 12, 6, 11, 2
b 4, 12, 7, 9, 5, 8, 11, 3, 6, 10, 2
c 2, 9, 4, 10, 7, 5, 3, 8, 12, 6, 11
d 6, 10, 7, 3, 9, 12, 8, 5, 11, 4, 2
e 11 – 88, 55, 22, 121, 66, 33, 110, 44, 77, 132, 99
4 – 32, 20, 8, 44, 24, 12, 40, 16, 28, 48, 36
6 – 48, 30, 12, 66, 36, 18, 60, 24, 42, 72, 54
3 – 24, 15, 6, 33, 18, 9, 30, 12, 21, 36, 27

3 a 32 b 42 c 6 d 6 e 77 f 9
g 27 h 48 i 8 j 28 k 12 l 9
m 36 n 36 o 8 p 12 q 132 r 21
s 5 t 12 u 10 v 54

4 a 7 b 4 c 9 d 55 e 6 f 12
g 3 h 36 i 11 j 3 k 6 l 10
m 3 n 4 o 11 p 54 q 4 r 12
s 20 t 12 u 6 v 12

5 a ×3, ÷2, ×4, ÷3, ÷4, ×7 b ×4, ×2, ÷5, ×4, ÷8, ÷2
c ÷5, ×4, ×2, ÷4, ÷2, ×6 d ×3, ÷5, ×8, ÷2, ÷12, ×8
e ÷2, ÷6, ×11, ×2, ÷4, ×5 f ×2, ÷4, ×9, ÷6, ×4, ÷3
g ÷5, ×10, ÷4, ÷5, ×6, ÷3 h ×3, ×2, ÷3, ×7, ÷2, ÷3

Page 27

Game

Page 28

1 a 80 b 120 c 250 d 190 e 340 f 010
g 150 h 470 i 500 j 780 k 230 l 860
m 520 n 370 o 950 p 280 q 490 r 800
s 550 t 910 u 330 v 670 w 420

2 a 380 b 56 c 970 d 71 e 63 f 450
g 79 h 210 i 880 j 30 k 14 l 590
m 2340 n 1560 o 5470 p 3110 q 4080 r 230
s 148 t 104 u 7100 v 1180 w 600

Page 29

1 a 16, 12, 20, 28, 36, 40 b 18, 36, 42, 60, 66
c 36, 33, 24, 9, 21, 15 d 35, 30, 45, 50, 55
e 49, 63, 56, 35, 21, 14, 70 f 55, 88, 22, 33, 66
g 80, 60, 90, 100, 30, 20 h 16, 40, 64, 56, 24, 88
i 99, 27, 36, 90, 72, 81, 63 j 24, 48, 60, 36, 120

2 a 111 b 17 c 45 d 28
e 42 f 125 g 19 h 18

Page 30

1 707 **2** 706 **3** 697 **4** 898 **5** 897 **6** 789
7 777 **8** 979 **9** 633 **10** 782 **11** 895 **12** 820
13 1254 **14** 1217 **15** 822 **16** 1210

Page 31

1 6799 **2** 2987 **3** 9899 **4** 6198
5 10 030 **6** 8941 **7** 9100 **8** 6470
9 6827 **10** 5973 **11** 4077 **12** 5302

Page 32

1 342 **2** 452 **3** 314 **4** 631 **5** 328 **6** 235
7 487 **8** 159 **9** 354 **10** 305 **11** 284 **12** 159

Page 33

1 5373 **2** 4220 **3** 4354 **4** 5251
5 4502 **6** 7316 **7** 2842 **8** 2857
9 3264 **10** 2364 **11** 6353 **12** 7455

Page 34

1 a 393 b 488 c 826 d 969 e 528 f 954
g 1144 h 925 i 2838 j 2758 k 2690 l 2376

2 a 952 b correct c 1545 d 438
e correct f correct g 1245 h 2148

Page 35

1 6942 **2** 16 760 **3** 6588 **4** 9020
5 6195 **6** 24 101 **7** 10 824 **8** 20 928
9 13 719 **10** 14 340 **11** 6338 **12** 15 664

VERY WELL DONE

Pages 36–37

1 a 115 b 216 c 225 d 119 e 144 f 148
g 118 h 134 i 153 j 194 k 167 l 107
m 163 n 106 o 142 p 116 q 258 r 244

2 a 72 rem. 1 b 126 rem. 4 c 56 rem. 2 d 131 rem. 3
e 65 rem. 4 f 264 rem. 1 g 82 rem. 3 h 58 rem. 3
i 174 rem. 4 j 123 rem. 2 k 116 rem. 3 l 118 rem. 1
m 181 rem. 1 n 93 rem. 2 o 126 rem. 2 p 153 rem. 3
q 196 rem. 2 r 207 rem. 1 s 97 rem. 1 t 45 rem. 1
u 69 rem. 1 v 116 rem. 3 w 254 rem. 2 x 81 rem. 4

Page 38

1 37 rem. 6 **2** 34 rem. 1 **3** 324 rem. 3 **4** 145 rem. 5
5 23 rem. 4 **6** 41 **7** 235 rem. 2 **8** 256 rem. 1
9 190 rem. 3 **10** 137 rem. 2 **11** 339 rem. 4 **12** 424 rem. 3
13 127 rem. 3 **14** 230 rem. 5 **15** 533 rem. 1 **16** 135 rem. 2
17 297 rem. 1 **18** 152 rem. 1 **19** 248 rem. 1 **20** 137 rem. 7
21 130 rem. 1 **22** 142 rem. 6 **23** 637 rem. 3 **24** 145 rem. 1
25 126 rem. 3

Pages 39–41

1 433 children **2** K1066 **3** 436 km
4 a 3834 km b 5112 km c 7668 km d 10 224 km
5 K191 **6** 778 litres
7 a K1060 b K1325 c K1855 d K2385
8 2693 kg **9** K195 **10** K2009
11 a 2232 km b 460 km
12 a K1032 b K1376 c K2064 d K2408
13 354 km
14 a 1399 people b 8893 people
15 K185 **16** 63 rows

Page 42

1 $\frac{3}{8}$, three eighths **2** $\frac{3}{6}$, three sixths **3** $\frac{1}{4}$, one quarter
4 $\frac{4}{7}$, four sevenths **5** $\frac{6}{10}$, six tenths **6** $\frac{5}{8}$, five eighths
7 $\frac{1}{4}$, one quarter **8** $\frac{5}{12}$, five twelfths **9** $\frac{1}{8}$, one eighth
10 $\frac{4}{9}$, four ninths **11** $\frac{1}{6}$, one sixth **12** $\frac{3}{10}$, three tenths
13 $\frac{3}{5}$, three fifths **14** $\frac{8}{15}$, eight fifteenths **15** $\frac{6}{11}$, six elevenths
16 $\frac{7}{10}$, seven tenths

Page 43

1 a $\frac{1}{2}+\frac{1}{4}+\frac{1}{4}=1$ b $\frac{1}{4}+\frac{1}{4}+\frac{1}{2}=1$
c $\frac{1}{4}+\frac{1}{8}+\frac{1}{8}+\frac{1}{4}+\frac{1}{8}+\frac{1}{8}=1$ d $\frac{1}{4}+\frac{1}{4}+\frac{1}{8}+\frac{1}{8}+\frac{1}{4}=1$
e $\frac{1}{2}+\frac{1}{8}+\frac{1}{8}+\frac{1}{4}=1$ f $\frac{1}{4}+\frac{1}{8}+\frac{1}{8}+\frac{1}{2}=1$
g $\frac{1}{8}+\frac{1}{2}+\frac{1}{4}+\frac{1}{8}=1$ h $\frac{1}{4}+\frac{1}{2}+\frac{1}{8}+\frac{1}{8}=1$

2 a more b exactly c more

Page 44

1 $\frac{1}{4}$ **2** $1\frac{1}{2}$ **3** $\frac{3}{8}$ **4** $1\frac{5}{6}$ **5** $\frac{3}{10}$ **6** $2\frac{1}{4}$
7 $1\frac{2}{3}$ **8** $\frac{1}{3}$ **9** $\frac{5}{12}$ **10** $3\frac{1}{3}$ **11** $\frac{7}{8}$ **12** $\frac{1}{6}$

Page 45

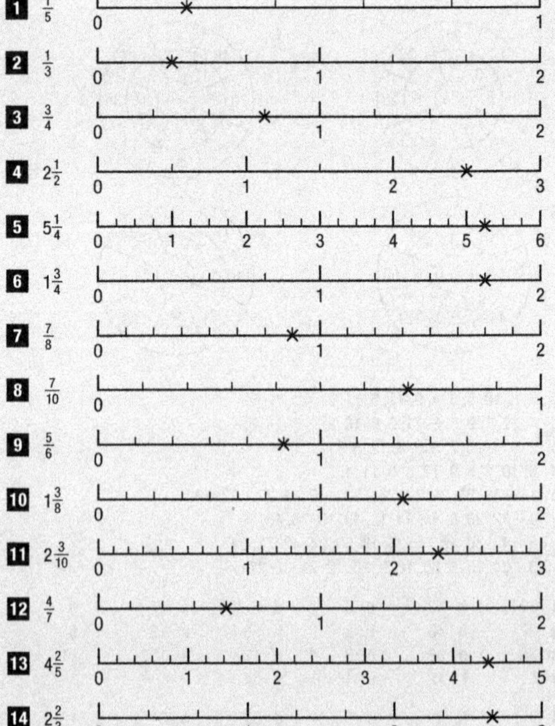

Pages 46–47

1 a $\frac{1}{4}$ b $\frac{1}{3}$ c $\frac{1}{7}$ d $\frac{1}{6}$ e $\frac{1}{2}$ f $\frac{1}{4}$
g $\frac{3}{4}$ h $\frac{5}{6}$ i $\frac{8}{9}$ j $\frac{2}{5}$ k $\frac{7}{8}$ l $\frac{4}{5}$

2 a $\frac{1}{10}$ b $\frac{1}{8}$ c $\frac{2}{10}$ d $\frac{2}{7}$ e $\frac{1}{7}$ f $\frac{1}{10}$
g $\frac{2}{3}$ h $\frac{2}{8}$ i $\frac{7}{9}$

3 a $\frac{2}{5}, \frac{1}{3}, \frac{3}{10}, \frac{1}{8}, \frac{4}{9}$ b $\frac{1}{3}, \frac{1}{2}, \frac{1}{8}, \frac{6}{9}$ c $\frac{1}{2}, \frac{2}{7}, \frac{5}{8}$

Page 48

Game

Page 49

1 one half **2** $\frac{1}{4}$ **3** $\frac{1}{8}$

4 one third **5** one fifth **6** $\frac{1}{10}$

Teacher to check all drawings of shapes, groups and number lines.

Page 50

1 $\frac{4}{5}$ **2** $\frac{7}{10}$ **3** $\frac{5}{8}$ **4** $\frac{4}{7}$ **5** $\frac{3}{9}$ **6** $\frac{7}{10}$

7 1 **8** $\frac{2}{6}$ **9** $\frac{5}{9}$ **10** $\frac{1}{3}$ **11** $\frac{3}{8}$ **12** $\frac{7}{11}$

13 $\frac{11}{12}$ **14** $\frac{2}{3}$ **15** $\frac{5}{7}$ **16** $\frac{1}{2}$ **17** $\frac{1}{5}$ **18** $\frac{3}{6}$

Page 51

1 0.7, $\frac{7}{10}$ **2** 0.9, $\frac{9}{10}$ **3** 0.6, $\frac{6}{10}$ **4** 0.5, $\frac{5}{10}$

5 0.2, $\frac{2}{10}$ **6** 0.8, $\frac{8}{10}$ **7** 0.3, $\frac{3}{10}$ **8** 0.1, $\frac{1}{10}$

Page 52

1 a $\frac{33}{100}$, 0.33 b $\frac{67}{100}$, 0.67 c $\frac{11}{100}$, 0.11
d $\frac{74}{100}$, 0.74 e $\frac{28}{100}$, 0.28 f $\frac{52}{100}$, 0.52

2 a $\frac{20}{100}, \frac{2}{10}$, 0.2 b $\frac{90}{100}, \frac{9}{10}$, 0.9 c $\frac{30}{100}, \frac{3}{10}$, 0.3
d $\frac{50}{100}, \frac{5}{10}$, 0.5 e $\frac{10}{100}, \frac{1}{10}$, 0.1 f $\frac{70}{100}, \frac{7}{10}$, 0.7

Pages 53–54

1 a 0.8 b 0.31 c 0.5 d 0.57 e 0.2 f 0.28
g 0.9 h 0.75 i 0.1 j 0.6 k 0.6 l 0.4
m 0.4 n 0.12 o 0.7 p 0.07 q 0.3 r 0.03
s 0.19 t 0.08 u 0.9 v 0.1 w 1.1 x 2.3
y 5.9 z 8.4

2 a 2.18 m b 7.53 m c 1.5 m d 4.8 m e 5.87 m f 3.2 m
g 1.45 m h 6.12 m i 4.86 m j 9.47 m k 0.84 m l 0.77 m
m 2.08 m n 8.03 m o 5.06 m

3 a K6.45 b K3.75 c K9.25 d K1.70 e K5.10 f K2.50
g K10.99 h K3.05 i K12.04 j K8.15 k K25.60 l K40.35
m K16.08 n K33.28 o K11.56

Pages 55–56

1 a 0.6 b 0.9 c 0.1 d 0.5 e 0.49 f 0.63
g 0.21 h 0.38 i 0.3 j 0.03 k 0.8 l 0.08
m 0.05 n 0.77 o 0.4 p 0.8 q 0.29 r 0.14
s 0.5 t 0.7 u 0.01 v 0.17 w 0.87 x 0.1

2 a 1.5 b 5.7 c 2.4 d 8.1 e 4.36 f 7.84
g 3.49 h 6.7 i 5.03 j 1.08 k 9.06 l 4.11
m 2.7 n 6.9 o 11.3 p 16.2 q 10.53 r 15.18
s 12.09 t 25.01 u 17.21 v 9.4 w 7.5 x 2.05

3 a $5\frac{67}{100}$ b $\frac{6}{10}$ c $1\frac{4}{100}$ d $8\frac{92}{100}$ e $\frac{7}{100}$ f $3\frac{5}{10}$
g $2\frac{77}{100}$ h $\frac{64}{100}$ i $4\frac{59}{100}$ j $6\frac{3}{100}$ k $\frac{2}{10}$ l $1\frac{81}{100}$
m $11\frac{8}{100}$ n $19\frac{1}{10}$ o $15\frac{25}{100}$ p $12\frac{9}{10}$

4 a 2 m 21 cm b 86 cm c 5 m 70 cm d 4 m 20 cm
e 90 cm f 1 m 25 cm g 8 m 50 cm h 3 m 75 cm
i 10 m 60 cm j 12 m 63 cm k 9 m 8 cm l 6 m 5 cm

Pages 56–57

1 a 2.75 b 32.5 c 5.6 d 7.89 e 1.3 f 15.2
g 23.45 h 0.7 i 0.81 j 12.5 k 3.6 l 10.2
m 8.9 n 6.5 o 9.15

2 a 9.7, 8.6, 7.9, 7.8, 7.6 b 5.1, 4.0, 3.7, 3.5, 2.4
c 3.5, 3.1, 2.9, 2.1, 2.0 d 5.3, 5.0, 4.9, 4.6, 4.3
e 8.6, 8.4, 8.1, 7.9, 7.8 f 1.9, 1.7, 1.1, 0.9, 0.6
g 6.8, 6.5, 6.1, 5.8, 5.5 h 9.6, 9.3, 9.0, 8.9, 8.6

3 a 6.07, 6.67, 6.7, 6.72, 6.79, 6.85 b 1.08, 1.36, 1.38, 1.41, 1.49, 1.8
c 2.01, 2.09, 2.1, 2.12, 2.19, 2.2 d 2.98, 3.09, 3.18, 3.5, 3.56, 3.9
e 4.09, 4.19, 4.89, 4.9, 4.95, 4.99 f 0.07, 0.17, 0.68, 0.7, 0.75, 0.77
g 8.06, 8.07, 8.56, 8.59, 8.6, 8.63 h 5.01, 5.05, 5.11, 5.15, 5.5, 5.51
i 7.02, 7.23, 7.29, 7.3, 7.32, 7.35 j 9.06, 9.59, 9.6, 9.65, 9.67, 9.69
k 0.04, 0.14, 0.39, 0.4, 0.41, 0.47 l 2.03, 2.16, 2.36, 2.39, 2.59, 2.6

Pages 58–59

1 a 4 m b 1 m c 6 m d 3 m e 2 m f 4 m
g 6 m h 2 m i 8 m j 2 m k 3 m l 6 m
m 9 m n 1 m o 2 m p 6 m q 8 m r 5 m
s 0 m t 1 m u 10 m v 13 m w 11 m x 16 m

2 a 6 b 2 c 8 d 6 e 4 f 11
g 6 h 0 i 2 j 8 k 4 l 10
m 5 n 1 o 4 p 3 q 11 r 9
s 1 t 6 u 15 v 13 w 19 x 12

3 a 12.8 m b 25.4 m c 15.3 m d 33.8 m e 28.2 m f 11.3 m
g 40.7 m h 10.4 m i 17.5 m j 38.9 m k 20.1 m l 36.8 m
m 23.1 m n 19.4 m o 55.9 m p 49.3 m q 36.4 m r 61.1 m
s 42.1 m t 53.6 m u 70.6 m v 45.4 m w 82.8 m x 75.2 m

4 a 5.7 kg b 3.2 kg c 0.7 kg d 7.3 kg e 1.9 kg f 6.6 kg
g 2.1 kg h 11.4 kg i 8.2 kg j 0.2 kg k 4.8 kg l 9.8 kg
m 2.5 kg n 5.9 kg o 10.4 kg p 6.1 kg q 12.1 kg r 8.8 kg
s 15.4 kg t 20.9 kg u 10.8 kg v 17.5 kg w 11.4 kg x 19.2 kg

Page 60

1 1.4 **2** 8.9 **3** 3.1 **4** 10.3 **5** 17.1 **6** 3.8

7 27.2 **8** 3.5 **9** 3.7 **10** 30.4 **11** 2.7 **12** 14.7

13 29.3 **14** 5.7 **15** 41.2 **16** 8.7

Page 61

1 9.53 **2** 5.15 **3** 9.55 **4** 7.36

5 3.93 **6** 14.33 **7** 5.16 **8** 9.23

9 15.53 **10** 5.92 **11** 10.13 **12** 4.16

13 1.19 **14** 10.53 **15** 3.73 **16** 5.86

The name of the shape is TRIANGLE.

Page 62

1 9.6 **2** 8.8 **3** 15.9 **4** 7.5 **5** 28.2

6 28.8 **7** 19.6 **8** 25.6 **9** 7.2 **10** 52.5

11 58.0 **12** 40.2 **13** 22.52 **14** 19.02 **15** 51.24

16 19.52 **17** 22.08 **18** 54.72 **19** 41.45 **20** 25.48

Page 63

1 **a** 2.2 **b** 1.3 **c** 2.4 **d** 1.7 **e** 3.4 **f** 2.7
g 9.5 **h** 3.8 **i** 2.53 **j** 3.16 **k** 1.84 **l** 4.77

2 **a** 6.2 **b** 4.8 **c** 18.4 **d** 8.9 **e** 3.85 **f** 5.13
g 8.57 **h** 2.31 **i** 1.87 **j** 7.28

Pages 64–65

1 1.5 km **2** 13.4 L

3 **a** 13.5 m **b** 8.1 m **c** 21.6 m

4 6.7 m

5 **a** K21.90 **b** K1.85

6 28.5 kg **7** 2.6 km **8** K15.40

Assessment

Pages 66–67

1 B **2** C **3** C **4** D **5** A **6** B

7 D **8** B **9** A **10** C **11** A **12** D

13 B **14** B **15** C **16** A

Pages 68–71

1 **a** 34 **b** 38 **c** 36 **d** 48

2 **a** 80 **b** 90 **c** 80 **d** 110 **e** 68 **f** 89
g 58 **h** 84

3 **a** 9 **b** 12 **c** 5 **d** 7 **e** 9 **f** 9
g 12 **h** 7 **i** 24 **j** 17 **k** 28 **l** 17
m 32 **n** 35 **o** 45 **p** 37

4 **a** 12 **b** 36 **c** 77 **d** 48 **e** 72 **f** 21
g 30 **h** 48 **i** 27 **j** 42 **k** 28 **l** 121

5 **a** 9 **b** 9 **c** 6 **d** 6 **e** 6 **f** 10
g 6 **h** 8 **i** 7 **j** 4 **k** 4 **l** 7

6 **a** 130 **b** 470 **c** 240 **d** 610 **e** 3520 **f** 1250
g 2680 **h** 5430

7 **a** 18, 30, 12, 24, 60 **b** 45, 36, 27, 63, 81 **c** 12, 40, 48, 16, 36, 28

8 **a** 997 **b** 661 **c** 8779 **d** 8414

9 **a** 552 **b** 169 **c** 5732 **d** 2277

10 **a** 464 **b** 3206 **c** 6250 **d** 8840

11 **a** 149 **b** 158 **c** 1567 **d** 393

12 **a** 123 rem. 2 **b** 151 rem. 2 **c** 635 rem. 6 **d** 1454 rem. 1

13 **a** K1427 **b** 4441 L **c** 3745 kg

Pages 72–73

1 **a** $\frac{5}{8}$ **b** $\frac{3}{10}$ **c** $\frac{5}{9}$ **d** $\frac{5}{6}$

2 **a** No **b** Yes **c** No

3 **a** $\frac{2}{3}$ **b** $\frac{7}{10}$ **c** $1\frac{3}{4}$

4 $\frac{1}{4}, \frac{1}{3}, \frac{2}{7}, \frac{1}{8}$

5 **a** $\frac{7}{9}$ **b** $\frac{9}{10}$ **c** $\frac{5}{6}$ **d** $\frac{5}{7}$

6 **a** $\frac{7}{10}$ **b** $\frac{5}{12}$ **c** $\frac{4}{9}$ **d** $\frac{3}{8}$

Pages 74–76

1 **a** 0.6, $\frac{6}{10}$ **b** 0.7, $\frac{7}{10}$ **c** 0.3, $\frac{3}{10}$

2 **a** 0.7 **b** 0.43 **c** 0.3 **d** 0.59 **e** 0.05 **f** 0.8

3 **a** 3.27 m **b** 2.41 m **c** 5.06 m **d** 1.5 m **e** 9.17 m **f** 0.78 m

4 **a** 0.4 **b** 0.8 **c** 0.61 **d** 0.44 **e** 0.3 **f** 8.7
g 6.33 **h** 3.05

5 **a** $\frac{8}{10}$ **b** $2\frac{13}{100}$ **c** $1\frac{9}{100}$ **d** $\frac{3}{100}$ **e** $4\frac{63}{100}$ **f** $11\frac{2}{10}$
g $3\frac{87}{100}$ **h** $5\frac{9}{10}$

6 **a** 0.8 **b** 5.6 **c** 2.4 **d** 6.79 **e** 1.5 **f** 8.1
g 3.6 **h** 7.2

7 **a** 5 **b** 2 **c** 1 **d** 9 **e** 2 **f** 3
g 10 **h** 3

8 **a** 5.8 **b** 2.1 **c** 4.6 **d** 8.2 **e** 0.4 **f** 1.8
g 7.9 **h** 3.1

9 **a** 12.5 **b** 25.3 **c** 12.21 **d** 12.84

10 **a** 4.8 **b** 3.7 **c** 4.59 **d** 3.54

11 **a** 16.8 **b** 99.2 **c** 26.85 **d** 67.55 **e** 4.7 **f** 6.3
g 13.6 **h** 22.7

12 **a** 7.2 kg **b** K1.75

Strand: Measurement

Page 77

1 10 cm 5 mm, 105 mm
2 12 cm, 120 mm
3 14 cm, 140 mm
4 16 cm 5 mm, 165 mm
5 20 cm, 200 mm
6 14 cm, 140 mm

Pages 78–79

1 a 300 cm b 500 cm c 200 cm d 800 cm e 150 cm f 450 cm g 650 cm h 950 cm i 540 cm j 360 cm k 170 cm l 730 cm m 225 cm n 165 cm o 412 cm p 78 cm q 341 cm r 572 cm s 56 cm t 187 cm u 807 cm v 1039 cm w 1260 cm x 1140 cm

2 a 2 m b 6 m c 1 m d 10 m e 15 m f 18 m g 11 m h 13 m i 3.5 m j 7.5 m k 4.5 m l 8.5 m m 2.7 m n 1.2 m o 0.9 m p 5.4 m q 3.25 m r 6.72 m s 1.17 m t 0.39 m u 8.43 m v 0.74 m w 12.56 m x 10.8 m

3 a 80 mm b 30 mm c 50 mm d 10 mm e 45 mm f 25 mm g 75 mm h 105 mm i 6 mm j 14 mm k 68 mm l 51 mm m 26 mm n 83 mm o 49 mm p 32 mm q 17 mm r 99 mm s 74 mm t 3 mm u 152 mm v 205 mm w 128 mm x 109 mm

4 a 6 cm b 9 cm c 15 cm d 4 cm e 12 cm f 20 cm g 17 cm h 3 cm i 4.5 cm j 8.5 cm k 10.5 cm l 21.5 cm m 3.8 cm n 2.1 cm o 5.7 cm p 9.4 cm q 36.4 cm r 20.8 cm s 63.5 cm t 47.6 cm u 105.5 cm v 130.2 cm w 116.9 cm x 157.9 cm

Page 80

1 16 cm **2** 22 cm **3** 15 cm **4** 22 cm
5 19 cm **6** 18 cm **7** 19 cm **8** 19 cm

Page 81

1 12 cm^2 **2** 22 cm^2 **3** 15 cm^2 **4** 12 cm^2 **5** 10 cm^2
6 16 cm^2 **7** 15 cm^2 **8** 24 cm^2 **9** 14 cm^2 **10** 20 cm^2

Pages 82–83

1 a 6000 mL b 3000 mL c 9000 mL d 2000 mL e 10 000 mL f 5500 mL g 1500 mL h 500 mL i 4250 mL j 7250 mL k 750 mL l 8750 mL m 2750 mL n 9500 mL o 6250 mL p 1750 mL q 250 mL r 3250 mL s 5750 mL t 10 500 mL

2 a 7 L b 1 L c 5 L d $\frac{1}{2}$ L or 500 mL e 4 L f 8 L 600 mL g 3 L 100 mL h 1 L 750 mL i 2 L 890 mL j 795 mL k 6 L 345 mL l 7 L 8 mL m 4 L 903 mL n 1 L 857 mL o 5 L 76 mL p 3 L 892 mL q 8 L 117 mL r 2 L 34 mL s 9 L 102 mL t 6 L 550 mL

3 1500 mL, $2\frac{1}{4}$ L, 1010 mL, $5\frac{1}{2}$ L, 2500 mL

4 $\frac{3}{4}$ L, 1 L, $1\frac{1}{4}$ L, 900 mL, $2\frac{1}{2}$ L, 505 mL

5 1 L, 900 mL, 1200 mL, $1\frac{1}{4}$ L, 1050 mL, 800 mL

6 2 L, 1800 mL, 1790 mL, 2030 mL

Pages 84–85

1 a 500 g b 250 g c 750 g d 1000 g e 2500 g f 1250 g g 2000 g h 1500 g i 1750 g j 100 g k 900 g l 300 g m 700 g n 1100 g o 2300 g

2 a 8 kg b 2 kg c 1 kg d 7 kg e 4 kg 500 g f 2 kg 800 g g 3 kg 500 g h 8 kg 100 g i 850 g j 1 kg 170 g k 5 kg 980 g l 6 kg 700 g m 9 kg 500 g n 8 kg 465 g o 1 kg 14 g p 2 kg 63 g q 6 kg 5 g r 5 kg 10 g s 3 kg 609 g t 1 kg 401 g

3 1300 g, $1\frac{1}{4}$ kg, $2\frac{1}{2}$ kg, 1050 g, 1250 g

4 750 g, 1000 g, $1\frac{1}{2}$ kg, 900 g, 670 g, $\frac{3}{4}$ kg

5 1000 g, $\frac{3}{4}$ kg, 1220 g, 1150 g

6 $1\frac{1}{2}$ kg, 2000 g, $1\frac{3}{4}$ kg, 1970 g, 2150 g

Page 86

1 8 o'clock, 8:00 **2** half past 5, 5:30 **3** 2 o'clock, 2:00
4 11 o'clock, 11:00 **5** half past 1, 1:30 **6** 4 o'clock, 4:00
7 half past 9, 9:30 **8** 7 o'clock, 7:00 **9** half past 8, 8:30
10 9 o'clock, 9:00 **11** half past 4, 4:30 **12** 5 o'clock, 5:00

Pages 87–88

1 a quarter past 7 b quarter past 11 c quarter past 1 d quarter to 4 e quarter to 7 f quarter to 12 g quarter past 5 h quarter to 3 i quarter past 9 j quarter to 10 k quarter past 8 l quarter to 6

2 a 11:15 b 3:15 c 5:45 d 8:45

3 a quarter past 12, 12:15 b quarter to 5, 4:45 c quarter past 5, 5:15 d quarter to 11. 10:45 e quarter past 7, 7:15 f quarter to 7, 6:45

4 a

b

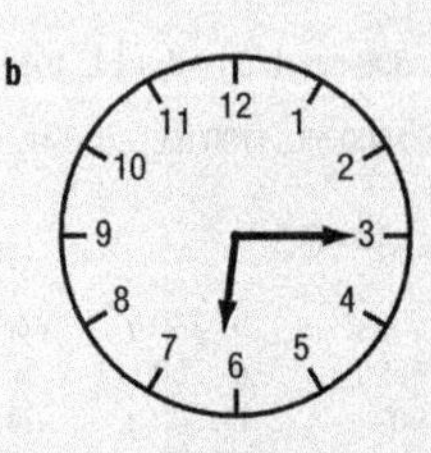

c

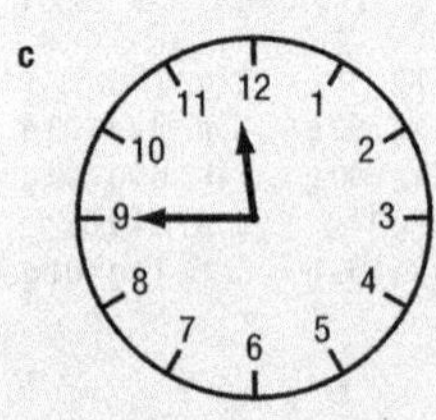

d

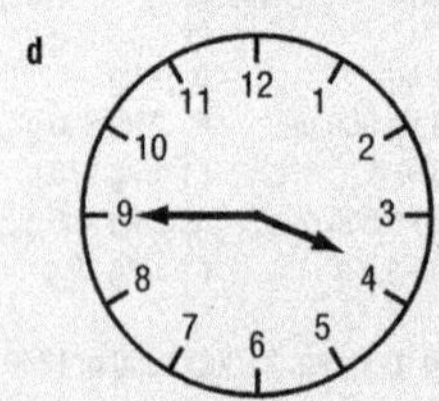

e

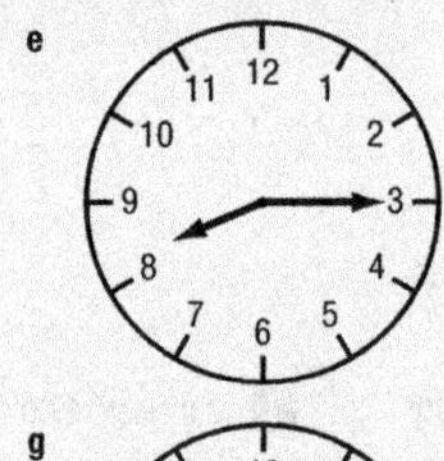

f

g

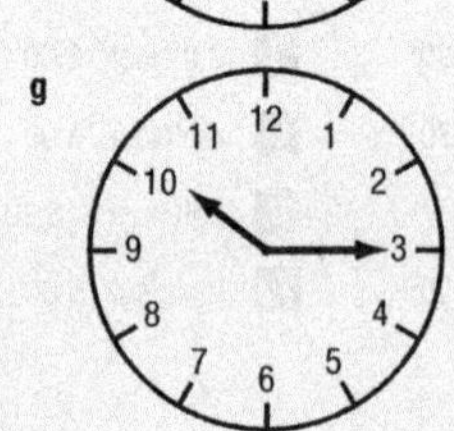

h

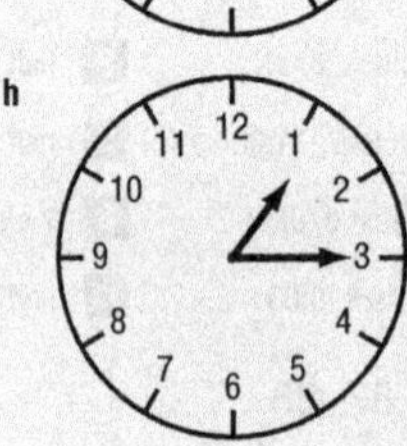

Page 89

1 a 12 b 60 c 120 d 48 e 96 f 24
g 84 h 36 i 72 j 18 k 54 l 30
m 90 n 66 o 114

2 a 14 b 70 c 28 d 35 e 56 f 7
g 21 h 63 i 77 j 49 k 84 l 42

3 a 60 b 300 c 120 d 600 e 180 f 480
g 360 h 540 i 420 j 90 k 330 l 210
m 270 n 150 o 510

4 a 48 b 96 c 24 d 192 e 120 f 240
g 72 h 168 i 144 j 36 k 132 l 228
m 84 n 156 o 60

Pages 90–91

1 a 7:30 b 5:30 c 4:30 d 7:00

2 a 3:45 b 4:30 c 5:15 d 1:15

3 a 7:45 b 7 o'clock c 8:30 d 8:15

4 a 11:15 b 10:30 c 11:45 d 11 o'clock

5 a 7:30 b 7 o'clock c 8:45 d 6:45

6 a 8:30 b 8:15 c 10 o'clock d 9:45

7 a 1:30 b 2:15 c 3:15 d 1:45

8 a 3 o'clock b 1:15 c 4:15 d 2:30

Pages 91–92

1 5:30 **2** 5:30 **3** 3:15 **4** 7 o'clock

5 3:45 **6** 5 o'clock **7** 12:30 **8** 4:45

Assessment

Pages 93–97

1 a B b E c C d D e C, D, B, A, E

2 a 700 cm b 250 cm c 450 cm d 390 cm
e 70 cm f 163 cm g 250 cm h 588 cm

3 a 8 m b 5.5 m c 14 m d 18.5 m
e 7.2 m f 2.9 m g 0.4 m h 1.13 m

4 a 70 mm b 20 mm c 35 mm d 65 mm
e 8 mm f 81 mm g 47 mm h 13 mm

5 a 3 cm b 11 cm c 7.5 cm d 25.5 cm
e 40 cm f 31.8 cm g 50.6 cm h 5.7 cm

6 a 18 cm b 18 cm

7 a 8 cm^2 b 16 cm^2

8 a 5000 mL b 2500 mL c 250 mL d 9 L
e 1.5 L f 6 L

9 a Yes b No c No d Yes
e Yes f No g No h No

10 a 1000 g b 500 g c 3250 g d 3 kg
e $7\frac{1}{2}$ kg f 5 kg

11 a Yes b No c Yes d Yes
e No f Yes g Yes h No

12 a quarter past 2, 2:15 b quarter past 6, 6:15
c quarter to 12, 11:45

13 a

b

c

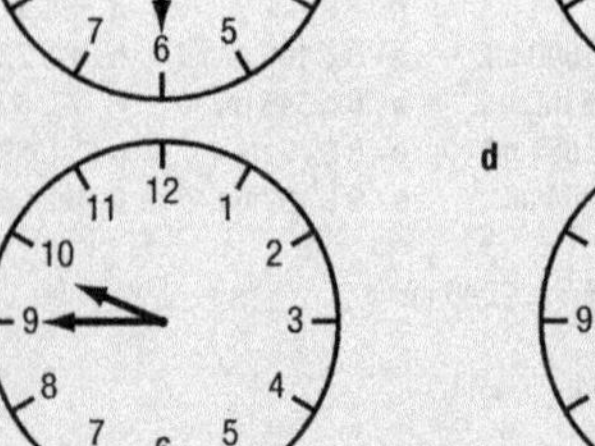

d

14 a 36 b 42 c 42 d 49
e 240 f 390 g 72 h 108

15 a 3:30 b 2 o'clock c 2:45 d 1:15

16 a 6:15 b 5:30 c 6:30 d 4:15

17 a 11:45 b 8:15

Strand: Space and Shape

Page 98

1 triangle **2** circle **3** square **4** cone **5** cube
6 pyramid **7** rectangle **8** trapezium **9** diamond **10** cylinder
11 sphere **12** pyramid **13** square **14** rectangle **15** triangle

Page 99

1 e **2** d **3** f **4** c **5** e **6** d **7** b

Page 100

1 a, d, f **2** b, c, e **3** a, e, f **4** a, b, d, e
5 d, e **6** b, c, f

Page 101

1 a, c, d, f **2** a, c, f **3** a, c, d **4** a, c, e
5 b, c, d **6** a, d

Page 102

1 Shape A = 4; Shape B = 3; Shape C = 1; Shape D = 1; Shape E = 2;
Shape F = 6; Shape G = 4; Shape H = 1; Shape I = 2
Teacher to check lines of symmetry.

2 Teacher to check drawn shapes.

Page 103

1 a, b, e and f **2** b, c, f and h **3** a, c, g and h

Page 104

Right angles = d, g and m; Straight angles = f, k and p; Acute angles = a, c, j and o;
Obtuse angles = b, i and n; Reflex angles = e, h and l

Page 105

1 b, a, e, c and d **2** e, b, a, d and c **3** c, a, e, d and b
4 d, b, c, e and a **5** d, a, e, b and c

Page 106

Game

Page 107

1 a F6 b C4 c A5 d E2 e E4 f B1
g C6 h E3 i C5 j F3 k D5 l D2
m A2 n F1 o D4 p E5 q C2 r E6

2 a 10t coin b pig c ball d clock
e arrow f boy g bucket h pencil
i dice j tin of Milo k frog l spoon

Page 108

1 South **2** East **3** West **4** South
5 East **6** North **7** West **8** East

Assessment

Pages 109–111

1 a rectangle b triangle c cylinder d cone

2 a any shape with four sides b any shape with five sides

3 a B b C c A

4 a 1 b 4 c 2

5 b and c

6 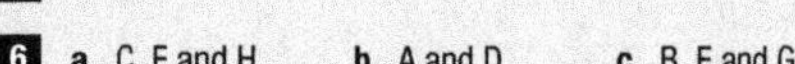
a C, F and H b A and D c B, E and G

7

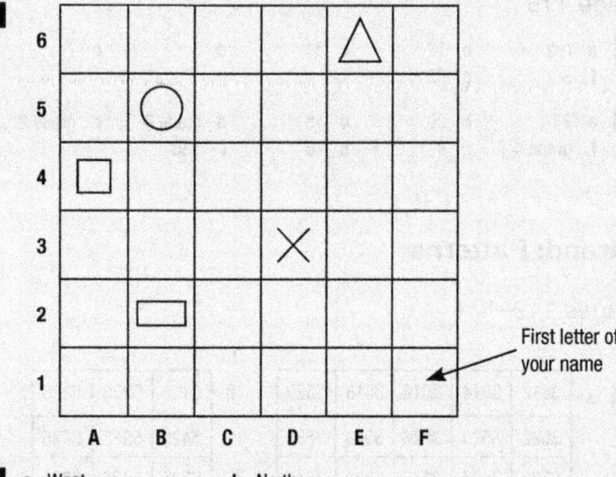

8 a West b North

Strand: Chance and Data

Page 112

1 a 4 b 5 c 2 d Teacher to check

2 a 2 b 3 c 10

3 a 10, 20 and 30 b 6 c 12

Pages 113–114

1 a Nora b Mary c Paul and Pia d 10 minutes e 12 minutes f 5 minutes g Albert h Charles i 8:30 j 8:05

2 a rice b 40 kg c 20 kg d 10 kg e 185 kg f 30 kg g K10 h 75t

Page 115

1 30 **2** 27 **3** Sunday **4** 69 **5** 18 **6** 15

7 Wednesday, Thursday, Friday and Saturday **8** Friday and Saturday

9 21 **10** 75

11 a 20 b 28 c 26

Pages 116–117

1 23 Feb **2** 1 Mar **3** 8 days **4** 8 days **5** 3

6 Papua New Guinea **7** 20 Mar **8** 4 days

9 Port Moresby to Townsville. Reasons given will vary.

10 14 May

Assessment

Page 118

1 a red b 12 c 5 d 3 e 7 f 6 g 52

2 a 24 b 20 c 18 d grade 6 e grade 2 f grade 4 g 4 h 46 i 130

Strand: Patterns

Pages 119–121

1 a

3512	3514	3516	3518	3520
3522	3524	3526	3528	3530
3532	3534	3536	3538	3540
3542	3544	3546	3548	3550

b

5125	5225	5325
5425	5525	5625
5725	5825	5925
6025	6125	6225

c

1841	1851	1861	1871	1881	1891	1901	1911	1921	1931
1941	1951	1961	1971	1981	1991	2001	2011	2021	2031
2041	2051	2061	2071	2081	2091	2101	2111	2121	2131
2141	2151	2161	2171	2181	2191	2201	2211	2221	2231
2241	2251	2261	2271	2281	2291	2301	2311	2321	2331
2341	2351	2361	2371	2381	2391	2401	2411	2421	2431

d

2067	2072	2077	2082
2087	2092	2097	2102
2107	2112	2117	2122
2127	2132	2137	2142
2147	2152	2157	2162
2167	2172	2177	2182
2187	2192	2197	2202
2207	2212	2217	2222
2227	2232	2237	2242

e

4763	4863	4963	5063
5163	5263	5363	5463
5563	5663	5763	5863
5963	6063	6163	6263
6363	6463	6563	6663
6763	6863	6963	7063
7163	7263	7363	7463
7563	7663	7763	7863
7963	8063	8163	8263

2 a 1682, 1782, 1882, 1982, 2082, 2182
b 4561, 4551, 4541, 4531, 4521, 4511
c 2091, 2096, 2101, 2106, 2111, 2116
d 7477, 7487, 7497, 7507, 7517, 7527
e 3214, 3114, 3014, 2914, 2814, 2714
f 1996, 1998, 2000, 2002, 2004, 2006
g 8113, 8108, 8103, 8098, 8093, 8088
h 5623, 5613, 5603, 5593, 5583, 5573
i 2758, 2858, 2958, 3058, 3158, 3258
j 3816, 3821, 3826, 3831, 3836, 3841
k 9553, 9551, 9549, 9547, 9545, 9543
l 4998, 4999, 5000, 5001, 5002, 5003
m 1985, 1995, 2005, 2015, 2025, 2035
n 5000, 4980, 4960, 4940, 4920, 4900
o 6890, 6940, 6990, 7040, 7090, 7140
p 7002, 7001, 7000, 6999, 6998, 6997
q 8459, 8479, 8499, 8519, 8539, 8559
r 2460, 2260, 2060, 1860, 1660, 1460
s 3210, 3160, 3110, 3060, 3010, 2960
t 1600, 2100, 2600, 3100, 3600, 4100
u 5422, 5322, 5222, 5122, 5022, 4922
v 4074, 4069, 4064, 4059, 4054, 4049
w 6975, 6995, 7015, 7035, 7055, 7075
x 9650, 9150, 8650, 8150, 7650, 7150

Pages 121–122

1 681, 671, 661, 651, 641; subtract 10

2 740, 780, 820, 860, 900; add 40

3 836, 841, 846, 851, 856; add 5

4 593, 590, 587, 584, 581; subtract 3

5 1300, 1320, 1340, 1360, 1380; add 20

6 3421, 3416, 3411, 3406, 3401; subtract 5

7 6999, 6997, 6995, 6993, 6991; subtract 2

8 5725, 5735, 5745, 5755, 5765; add 10

9 2692, 2699, 2706, 2713, 2720; add 7

10 955, 935, 915, 895, 875; subtract 20

11 8210, 8310, 8410, 8510, 8610; add 100

12 4110, 4060, 4010, 3960, 3910; subtract 50

13 14, 14.5, 15, 15.5, 16; add 0.5

14 24.5, 27, 29.5, 32, 34.5; add 2.5

15 17.5, 16, 14.5, 13, 11.5; subtract 1.5

16 8.75, 7.5, 6.25, 5, 3.75; subtract 1.25

17 37.25, 37.5, 37.75, 38, 38.25; add 0.25

18 $13\frac{1}{2}$, 13, $12\frac{1}{2}$, 12, $11\frac{1}{2}$; subtract $\frac{1}{2}$

19 $6\frac{1}{4}$, $6\frac{1}{2}$, $6\frac{3}{4}$, 7, $7\frac{1}{4}$; add $\frac{1}{4}$

20 $6\frac{1}{4}$, $5\frac{1}{2}$, $4\frac{3}{4}$, 4, $3\frac{1}{4}$; subtract $\frac{3}{4}$

21 $12\frac{1}{2}$, 11, $9\frac{1}{2}$, 8, $6\frac{1}{2}$; subtract $1\frac{1}{2}$

22 $43\frac{1}{4}$, $45\frac{1}{2}$, $47\frac{3}{4}$, 50, $52\frac{1}{4}$; add $2\frac{1}{4}$

Page 123

1 g **2** f **3** j **4** l **5** k **6** i

7 a **8** d **9** c **10** h **11** b **12** e

Pages 124–125

1
- **a** 1500, 1507, 1514, 1521, 1528, 1535, 1542, 1549
- **b** 850, 842, 834, 826, 818, 810, 802, 794
- **c** 35, $34\frac{1}{2}$, 34, $33\frac{1}{2}$, 33, $32\frac{1}{2}$, 32, $31\frac{1}{2}$
- **d** 2345, 2355, 2365, 2375, 2385, 2395, 2405, 2415
- **e** 7233, 7228, 7223, 7218, 7213, 7208, 7203, 7198
- **f** 4305, 4205, 4105, 4005, 3905, 3805, 3705, 3605
- **g** 22, 24.5, 27, 29.5, 32, 34.5, 37, 39.5
- **h** 17, 17.5, 18, 18.5, 19, 19.5, 20, 20.5
- **i** 2, 4, 8, 16, 32, 64, 128, 256
- **j** 6400, 3200, 1600, 800, 400, 200, 100, 50
- **k** 3124, 3114, 3104, 3094, 3084, 3074, 3064, 3054
- **l** 5870, 5890, 5910, 5930, 5950, 5970, 5990, 6010
- **m** 1598, 1698, 1798, 1898, 1998, 2098, 2198, 2298
- **n** 2446, 2442, 2438, 2434, 2430, 2426, 2422, 2418
- **o** 50, $47\frac{1}{2}$, 45, $42\frac{1}{2}$, 40, $37\frac{1}{2}$, 35, $32\frac{1}{2}$
- **p** 40, $41\frac{1}{4}$, $42\frac{1}{2}$, $43\frac{3}{4}$, 45, $46\frac{1}{4}$, $47\frac{1}{2}$, $48\frac{3}{4}$
- **q** 36, 37.5, 39, 40.5, 42, 43.5, 45, 46.5
- **r** 15, 14.75, 14.5, 14.25, 14, 13.75, 13.5, 13.25
- **s** 6755, 6705, 6655, 6605, 6555, 6505, 6455, 6405
- **t** 3800, 3825, 3850, 3875, 3900, 3925, 3950, 3975
- **u** 1.5, 3, 6, 12, 24, 48, 96, 192
- **v** $\frac{1}{4}$, $\frac{1}{2}$, 1, 2, 4, 8, 16, 32
- **w** 70, 65.5, 61, 56.5, 52, 47.5, 43, 38.5
- **x** 28, $28\frac{3}{4}$, $29\frac{1}{2}$, $30\frac{1}{4}$, 31, $31\frac{3}{4}$, $32\frac{1}{2}$, $33\frac{1}{4}$
- **y** 4976, 4981, 4986, 4991, 4996, 5001, 5006, 5011
- **z** 7150, 7125, 7100, 7075, 7050, 7025, 7000, 6975

2
- **a** subtract 50 **b** subtract 2.5 **c** add 7.5 **d** multiply by 2
- **e** add $1\frac{1}{2}$ **f** divide by 2 **g** subtract 7 **h** add 15
- **i** subtract 20 **j** add 200 **k** add 1.5 **l** subtract 0.5
- **m** subtract $\frac{1}{4}$ **n** add 500 **o** subtract 150 **p** add 250

Assessment

Pages 126–127

1 **a**

5881	5886	5891	5896
5901	5906	5911	5916
5921	5926	5931	5936
5941	5946	5951	5956

b

3736	3836	3936	4036
4136	4236	4336	4436
4536	4636	4736	4836
4936	5036	5136	5236

2
- **a** 3154, 3134, 3114, 3094, 3074, 3054, 3034, 3014, 2994
- **b** 1869, 1919, 1969, 2019, 2069, 2119, 2169, 2219, 2269
- **c** 75, 73.5, 72, 70.5, 69, 67.5, 66, 64.5, 63

3
- **a** 496, 500, 504, 508, 512, 516; add 4
- **b** 563, 553, 543, 533, 523, 513; subtract 10
- **c** 7894, 7844, 7794, 7744, 7694; subtract 50
- **d** 5887, 5907, 5927, 5947, 5967; add 20

4 **a** add 3.5 **b** subtract $1\frac{1}{2}$ **c** add $\frac{3}{4}$

5
- **a** 1250, 1225, 1200, 1175, 1150, 1125, 1100, 1075
- **b** 60, 62.5, 65, 67.5, 70, 72.5, 75, 77.5
- **c** 4816, 4809, 4802, 4795, 4788, 4781, 4774, 4767

OXFORD
UNIVERSITY PRESS

253 Normanby Road, South Melbourne, Victoria 3205, Australia

Oxford University Press is a department of the University of Oxford. It furthers the University's objective of excellence in research, scholarship, and education by publishing worldwide in

Oxford New York

Auckland Cape Town Dar es Salaam Hong Kong Karachi Kuala Lumpur Madrid Melbourne Mexico City Nairobi New Delhi Shanghai Taipei Toronto

With offices in

Argentina Austria Brazil Chile Czech Republic France Greece Guatemala Hungary Italy Japan Poland Portugal Singapore South Korea Switzerland Thailand Turkey Ukraine Vietnam

OXFORD is a trade mark of Oxford University Press in the UK and in certain other countries

First published 2011
Reprinted 2018(D)

ISBN 978 0 19 557400 5

Typeset by Palmer Higgs
Printed and bound in Australia by Ligare Book Printers Pty Ltd